P9-DTU-616

CONCRETE IN HOT ENVIRONMENTS

Modern Concrete Technology Series

Series Editors

Arnon Bentur
National Building Research Institute
Technion-Israel Institute of Technology
Technion City
Haifa 32 000
Israel

Sidney Mindess
Department of Civil Engineering
University of British Columbia
2324 Main Mall
Vancouver
British Columbia
Canada V6T 1W5

Fibre Reinforced Cementitious Composites
A. Bentur and S. Mindess

Concrete in the Marine Environment
P.K. Mehta

Concrete in Hot Environments
I. Soroka

Durability of Concrete in Cold Climates
M. Pigeon and R. Pleau
(forthcoming)

High Strength Concrete
P.C. Aitcin
(forthcoming)

Concrete in Hot Environments

I. SOROKA

National Building Research Institute,
Faculty of Civil Engineering,
Technion — Israel Institute of Technology, Haifa, Israel

E & FN SPON

An Imprint of Chapman & Hall

London · Glasgow · New York · Tokyo · Melbourne · Madras

Published by E & FN Spon, an imprint of Chapman & Hall, 2–6 Boundary Row, London SE1 8HN, UK

Chapman & Hall, 2–6 Boundary Row, London SE1 8HN, UK

Blackie Academic & Professional, Wester Cleddens Road, Bishopbriggs, Glasgow G64 2NZ, UK

Chapman & Hall Inc., 29 West 35th Street, New York NY10001, USA

Chapman & Hall Japan, Thomson Publishing Japan, Hirakawacho Nemoto Building, 6F, 1-7-11 Hirakawa-cho, Chiyoda-ku, Tokyo 102, Japan

Chapman & Hall Australia, Thomas Nelson Australia, 102 Dodds Street, South Melbourne, Victoria 3205, Australia

Chapman & Hall India, R. Seshadri, 32 Second Main Road, CIT East, Madras 600 035, India

---

First edition 1993

Typeset in 12/14pt Bembo by Alden Multimedia, Northampton
Printed in Great Britain by the Alden Press, Oxford

ISBN 0 419 15970 3

A catalogue record for this book is available from the British Library

Library of Congress Cataloging-in-Publication data

Soroka, I. (Itzhak)
Concrete in hot environments / I. Soroka.
p. cm.—(Modern concrete technology series)
Includes bibliographical references and indexes.
ISBN 0 419 15970 3
1. Concrete construction—Hot weather conditions. 2. Concrete—Hot weather conditions. 3. Portland cement—Hot weather conditions. I. Title. II. Series.
TA682.48.S67 1993
620.1'3617—dc20

∞ Printed on permanent acid-free text paper, manufactured in accordance with ANSI/NISO Z39.48-1992 and ANSI/NISO Z39.48-1984 (Permanence of Paper).

To the future generation,
to **Or**, **Barak**, **Shir** and **Isar**

Foreword

Plain concrete is a brittle material, with low tensile strength and strain capacities. Nonetheless, with appropriate modifications to the material, and with appropriate design and construction methodologies, it is being used in increasingly sophisticated applications. If properly designed, concrete structures can be produced to be durable over a wide range of environmental conditions, including hot and cold climates, as well as aggressive exposure conditions such as in marine and highly polluted industrial zones. Indeed, our understanding of cementitious systems has advanced to the point where these systems can often be 'tailored' for various applications where ordinary concretes are limited.

However, the results of the current research, which make these advances possible, are still either widely scattered in the journal literature, or mentioned only briefly in standard textbooks. Thus, they are often unavailable to the busy engineering professional. The purpose of the *Modern Concrete Technology Series* is to provide a seies of volumes that each deal with a single topic of interest in some depth. Eventually, they will form a library of reference books covering all the major topics in modern concrete technology.

Recent advances in concrete technology have been obtained using the traditional materials science approach:

(1) characterisation of the microstructure;
(2) relationships between the microstructure and engineering properties;
(3) relationships between the microstructural development and the processing techniques; and
(4) selection of materials and processing methods to achieve composites with the desired characteristics.

Accordingly, each book in the series will cover both the fundamental scientific principles, and the practical applications. Topics will be discussed in terms of the basic principles governing the behaviour of the various cement composites, thus providing the reader with information valuable for engineering design and construction, as well as a proper background for assessing future developments.

The series will be of interest to practitioners involved in modern concrete technology, and will also be of use to academics, researchers, graduate students, and senior undergraduate students.

Concrete in Hot Environments, by Professor I. Soroka, is an additional book in this series, which focuses on the underlying processes governing the behaviour of concrete in hot climates. On this basis it provides guidelines for proper use and design of concrete exposed to such environmental conditions.

Arnon Bentur
Sidney Mindess

Preface

The specific problems associated with concrete and concreting in hot environments have been recognised for some decades. This recognition has manifested itself over the years at a few symposia and in hundreds of papers where relevant research results and field observations were presented and discussed. In other publications the practical conclusions from these available data and experiences have been summarised in the form of guidelines for hot climate concreting. This book is not intended as one more guide, but mainly to explain the influence of hot environments on the properties and behaviour of concrete, and to point out its practical implications. However, in order to understand these effects, basic knowledge of cement paste and concrete is essential. Although the author could have assumed that the reader either possesses the required knowledge or, when necessary, will consult other sources, he preferred to include, as far as possible, all the relevant information in the book. Accordingly, sections of the book discuss cement and concrete in general, but the discussion is confined only to those aspects which are relevant to the specific effects of hot environments. It is believed that such a presentation makes it much easier for the reader to follow and understand the discussion, and therefore it was adopted in this book.

I. Soroka

Acknowledgements

The book was written as part of the author's activity at the National Building Research Institute, Faculty of Civil Engineering, Technion—Israel Institute of Technology, Haifa, Israel. Over the years, a substantial body of experimental data and practical experience related to concrete in hot environments, has accumulated at the Institute. The author is indebted to his colleagues for making these data available and for allowing him to draw on their practical experience. Also to be acknowledged is the secretarial staff of the Institute for their devoted help and efforts in typing and producing the manuscript. Special thanks are due to Mrs Tamar Orell for her professional production of the artwork.

Part of the literature survey, which was required for writing this book, was carried out when the author, on Sabbatical leave from the Technion, spent the summer of 1990 at the Building Research Establishment (BRE), Garston, Watford, UK. The author is grateful to the Director of the BRE and his staff for their kind help and hospitality.

The book includes numerous figures and tables originally published by others elsewhere. The author is indebted to the relevant institutions, journals, etc. for permission to reproduce the following figures and tables:

The American Ceramic Society
735 Ceramic Place, Westerville, OH 43081-8720, USA (Fig. 1.3).

American Chemical Society
1155 Sixteenth St. NW, Washington, DC 20036, USA (Fig. 1.1).

American Concrete Institute (ACI)
PO Box 19150, 22400 West Seven Mile Road, Detroit, MI 48219, USA (Figs 1.4, 1.5, 2.13, 2.15, 2.16, 3.1, 3.4, 3.6, 3.12, 4.2, 4.6, 4.9, 4.16, 4.19, 4.20, 4.22, 4.23, 5.11, 6.11, 6.17, 7.15,

7.16, 8.14, 9.3, 9.13, 10.9, 10.19, and 10.20, and Tables 1.4, 9.1, and 9.2).

American Society of Civil Engineers
345 East 47th Street, New York, NY 10017-2398, USA (Fig. 3.7).

American Society for Testing and Materials (ASTM)
1916 Race St., Philadelphia, PA 19103-1187, USA (Figs 1.7, 3.10, 3.16, 4.11, 4.12, 7.17, 8.3, 9.8, 9.15 and 10.23, and Table 3.4).

Association Technique de l'Industrie des Liants Hydrauliques
8 Rue Villiot, 75012 Paris, France (Fig. 2.14).

The Bahrain Society of Engineers
PO Box 835, Manama, Bahrain (Fig. 10.14).

Beton Verlag
Postfach 110134, 4000 Dusseldorff 11 (Oberkassel), Germany (Figs 9.11 and 9.12).

British Cement Association
Wexham Springs, Slough, UK, SL3 6PL (Figs 6.9, 7.12, 8.5 and 8.10).

British Standard Institution
Linford Wood, Milton Keynes, UK, MK14 6LE (Figs 7.6 and 8.4, and Tables 10.1 and 10.2)

Bureau of Reclamation US Department of the Interior
PO Box 25007, Building 67, Denver Federal Center, Denver, CO 80225-0007, USA (Figs 1.6 and 4.3).

The Cement Association of Japan
17–33 Toshima, 4-chome, Kita-ku, Tokyo 114, Japan (Figs 2.9, 2.10, 6.16 and 7.5).

Il Cemento
Via Santa Teresa 23, 00198 Roma, Italy (Fig. 3.5).

Cement och Betong Institutet
S100-44 Stockholm, Sweden (Figs 10.17 and 10.18)

Chapman & Hall
2–6 Boundary Row, London, UK, SE1 8HN (Table 10.3).

Commonwealth Scientific and Industrial Research Organisation (CSIRO)
372 Albert St., East Melbourne, Victoria 3002, Australia (Figs 2.7 and 6.5).

Concrete Institute of Australia
25 Berry St., North Sydney, NSW 2060, Australia (Fig. 7.4).

Concrete Society
Framewood Road, Wexham, Slough, UK, SL3 6PJ (Fig. 8.15).

Elsevier Sequioa SA
Avenue de la Gare 50, 1003 Lausanne 1, Switzerland (Fig. 7.3).

EMPA
Uberlandstrasse 129, CH 8600 Dubendorf, Switzerland (Fig. 7.13).

Gauthier Villars
15, Rue Gossin, 92543 Montrouge Cedex, France (Fig. 8.12).

Instituto Eduardo Torroja de la Construction y del Cemento
Serrano Galivache s/n 28033, Madrid, Aptdo 19002, 28080 Madrid, Spain (Figs 5.5, 5.6 and 5.9).

The Macmillan Press Ltd
Houndmills, Basingstoke, Hampshire, UK, RG21 2XS (Figs 2.1, 6.1, 6.3, 6.6, 8.1 and 8.2)

National Building Research Institute, Faculty of Civil Engineering, Technion—Israel Institute of Technology
Technion City, Haifa 32000, Israel (Figs 3.11, 3.17, 5.4, 5.7, 5.8, 6.12, 6.13, 6.14, 6.15, 7.7, 7.8, 7.9, 8.6, 8.8, 8.9, 10.6, 10.8, 10.10, 10.12, 10.21 and 10.22).

National Bureau of Standards and Technology, US Department of Commerce
Gaithersburg, MD 20899, USA (Figs 2.5 and 7.11).

Pergamon Press
Headington Hill Hall, Oxford, UK, OX3 0BW (Figs 2.11, 3.3, 3.8, 5.3, 7.14, 9.2, 10.13 and 10.16).

Purdue University, School of Engineering
West Lafayette, IN 49907, USA (Fig. 9.14).

Rhelogical Acta, Dr. Dietrich Steinkoptf Verlag
6100 Darmstadt, Saalbaustrasse 12, Germany (Fig. 8.13).

RILEM Materials & Structures
Pavillon due CROUS, 61 av. du Pdt Wilson, 94235 Cachan Cedex, France (Figs 3.9, 5.10 and 8.12)

Sindicato Nacional da Industria do Cimento
Rua da Assembleia no. 10 grupo 4001, CEP 2001, Rio de Janeiro, RJ, Brazil (Fig. 9.4).

Stuvo/VNC—The Netherlands
Postbus 3011, 5203 DA's Hertogenbosch, The Netherlands (Figs 3.14, 9.10 and 10.15, and Table 9.3).

Technical Research Centre of Finland
PO Box 26 (Kemistintie 3), SF-02151 Espoo, Finland (Fig. 8.11).

Thomas Telford Publications
Thomas Telford House, 1 Heron Quay, London, UK, E14 4JD (Figs 3.13, 4.1, 6.8 and 8.7).

Transportation Research Board, National Research Council
2101 Constitution Ave., Washington, DC 20418, USA (Fig. 4.21).

Universitat Hannover, Institut fur Baustoffkunde und Materialprufung
Nienburges Strasse 3, D-3000 Hannover, Germany (Figs 5.2, 10.5 and 10.7).

University of Toronto Press
10 St. Mary St., Suite 700, Toronto, Ontario, Canada, M4Y 2W8 (Figs 1.8, 9.5 and 9.6).

Zement–Kalk–Gips, Bauverlag GmbH
Postfach 1460, D-6200 Wiesbaden, Germany (Figs 2.8 and 9.9).

The author is also grateful to the authors of the papers from which the figures and tables were reproduced. Direct reference to them is made in the appropriate places.

Contents

2 Setting and Hardening

3 Mineral Admixtures and Blended Cements

Chapter 1
Portland Cement

1.1. INTRODUCTION

Portland cement is an active hydraulic binder, i.e. a 'binder that sets and hardens by chemical interaction with water and is capable of doing so under water without the addition of an activator such as lime' (BS 6100, section 6.1, 1984). It is obtained by burning, at a clinkering temperature (about 1450°C), a homogeneous predetermined mixture of materials comprising lime (CaO), silica (SiO_2), a small proportion of alumina (Al_2O_3), and generally iron oxide (Fe_2O_3). The resulting clinker is finely ground (i.e. average particle size of 10 μm) together with a few percent of gypsum to give, what is commonly known as, Portland cement. This is, however, a generic term for various forms (types) of Portland cement which include, in addition to ordinary Portland cement (OPC), rapid-hardening Portland cement (RHPC), low-heat Portland cement (LHPC), sulphate-resisting Portland cement (SRPC) and several others. It will be shown later that the different forms of the cement are produced by changing the proportions of the raw materials, and thereby, also, the mineralogical composition of the resulting cements (see section 1.5).

1.2. MAJOR CONSTITUENTS

Cement is a heterogeneous material made up of several fine-grained minerals which are formed during the clinkering process. Four minerals, namely Alite, Belite, Celite and a calcium-aluminate phase, make up some 90% of the cement and are collectively known, therefore, as 'major constituents'. Accordingly, the remaining 10% are known as 'minor constituents'.

The structure of the cement constituents is not always exactly known and in engineering applications their composition is usually written, therefore, in a simple way as made up of oxides, i.e. in a form which, although representing their chemical composition, does not imply any specific structure. For example, the composition of the Alite, which is essentially tricalcium silicate, is written as $3CaO.SiO_2$. Moreover, in cement chemistry it is usual to describe each oxide by a single letter, namely, $CaO = C$, $SiO_2 = S$, $Al_2O_3 = A$, $Fe_2O_3 = F$ and $H_2O = H$. Accordingly, the tricalcium silicate is written as C_3S.

The properties of Portland cement are determined qualitatively, but not necessarily quantitatively, by the properties of its individual constituents and their content in the cement. Hence, the following discussion deals, in the first instance, with the properties of the individual constituents, whereas the properties of the cement, with respect to its composition, are dealt with later in the text.

1.2.1. Alite

Alite is essentially tricalcium silicate, i.e. $3CaO.SiO_2$ or C_3S. Its content in OPC is about 45%, and due to this high content, the properties and behaviour of the latter are very similar to those of Alite. Alite as such is a hydraulic binder. On addition of water, hydration takes place bringing about setting and subsequent hardening in a few hours. If not allowed to dry, the resulting solid gains strength with time mainly during the first 7–10 days. The compressive strength of the set Alite is comparatively high, ultimately reaching a few tens of MPa (Fig. 1.1). The hydration of the Alite, similar to the hydration of the other constituents of the cement, is exothermic with the quantity of heat liberated (i.e. the heat of hydration) being about 500 J/g.

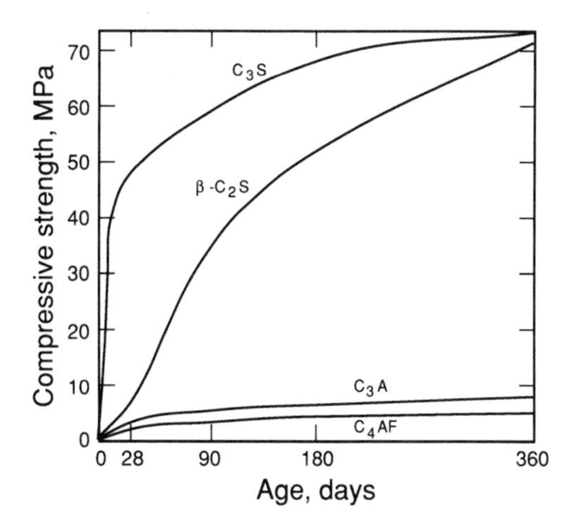

Fig. 1.1. Compressive strength of major constituents of Portland cement. (Adapted from Ref. 1.1).

1.2.2. Belite

Belite in Portland cement is essentially dicalcium silicate, i.e. $2CaO.SiO_2$ or C_2S. That is, a Belite is a calcium silicate with a poorer lime content as compared with Alite. Its average content in OPC is about 25%.

On addition of water the Belite hydrates liberating a comparatively small quantity of heat, i.e. about 250 J/g. Belite hydrates slowly and setting may take a few days. Strength development is also slow and, provided enough moisture is available, continues for weeks and months. Its ultimate strength, however, is rather high being of the same order as that of the Alite (Fig. 1.1).

1.2.3. Tricalcium Aluminate

In its pure form tricalcium aluminate ($3CaO.Al_2O_3$ or C_3A) reacts with water almost instantaneously and is characterised by a flash set which is accompanied by a large quantity of heat evolution, i.e. about 850 J/g. In moist air most of the strength is gained within a day or two, but the strength, as such, is rather low (Fig. 1.1). In water the set C_3A paste disintegrates, and C_3A may not be regarded, therefore, as a hydraulic binder. Its average content in OPC is about 10%. It will be seen later that the presence of C_3A makes Portland cement vulnerable to sulphate attack (see section 1.5.3).

1.2.4. Celite

Celite is the iron-bearing phase of the cement and it is, therefore, sometimes referred to as the ferrite phase. Celite is assumed to have the average composi-

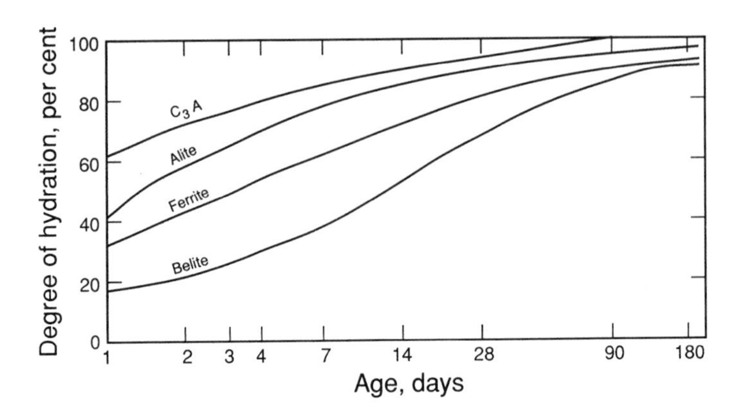

Fig. 1.2. Hydration of Portland cement constituents with time. (Data taken from Ref. 1.2).

tion of tetracalcium aluminoferrite ($4CaO.Al_2O_3.Fe_2O_3$ or C_4AF) and its average content in OPC is about 8%.

The Celite hydrates rapidly and setting occurs within minutes. The heat evolution on hydration is approximately 420 J/g. The development of strength is rapid but ultimate strength is rather low (Fig. 1.1). Celite imparts to the cement its characteristic grey colour, i.e. in the absence of the latter phase the colour of cement is white.

1.2.5. Summary

The different properties of the four major cement constituents are summarised in Figs 1.1 and 1.2, and in Table 1.1. It may be noted (e.g. Fig. 1.1) that the compressive strength of both calcium silicates (i.e. C_2S and C_3S) is much higher than the strengths of the C_3A and the C_4AF. It can also be noted that the ultimate strengths of C_2S and the C_3S are essentially the same, but the rate of strength development of the C_3S is higher than that of the C_2S. The considerable differences in the rates of hydration of the different constituents are reflected in Fig. 1.2. It can be seen that after 24 h approximately 65% of the C_3A hydrated as compared to only 15% of the C_2S. Additional differences may be noted in some other properties such as the rate of setting, the heat of hydration, etc. It will be seen later that all these differences are utilised to produce cements of different properties, i.e. to produce different types of Portland cement (see section 1.5).

Table 1.1. Properties of the Major Constituents of Portland Cement

	Alite	*Belite*	*Aluminate Phase*	*Celite*
Approximate chemical composition	Tricalcium silicate $3CaO.SiO_2 (C_3S)$	Dicalcium silicate $2CaO.SiO_2 (C_2S)$	Tricalcium aluminate $3CaO.Al_2O_3 (C_3A)$	Tetracalcium aluminoferrite $4CaO.Fe_2O_3.Al_2O_3 (C_4AF)$
Setting	Rapid (hours)	Slow (days)	Instantaneous	Very rapid (minutes)
Strength development	Rapid (days)	Slow (weeks)	Very rapid (1 day)	Very rapid (1 day)
Ultimate strength	High: 10s of MPa	Probably high: 10s of MPa	Low: a few MPa	Low: a few MPa
Heat of hydration	Medium: $\sim$500 J/g	Low: $\sim$250 J/g	Very high: $\sim$850 J/g	Medium: $\sim$420 J/g
Remarks	Characteristic constituent of Portland cements		Unstable in water, vulnerable to sulphate attack	Imparts to the cement its characteristic grey colour

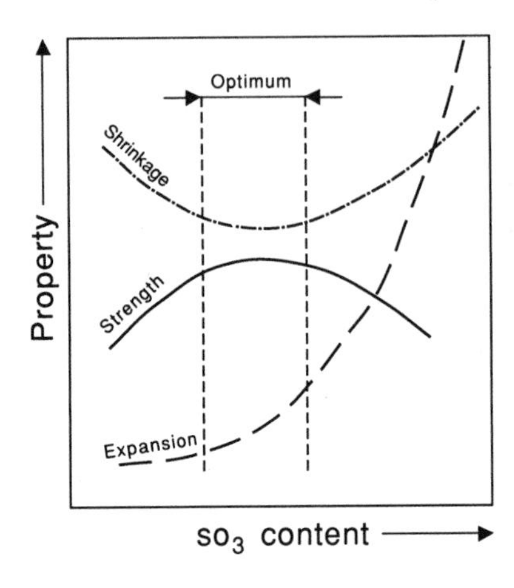

Fig. 1.3. Schematic description of optimum gypsum content.

1.3. MINOR CONSTITUENTS

1.3.1. Gypsum ($CaSO_4 \cdot 2H_2O$)

It was pointed out earlier (section 1.2.3) that the C_3A reacts with water almost instantaneously, bringing about an immediate stiffening of its paste. In OPC the C_3A content is about 10%, and this content is high enough to produce flash set. In order to avoid this, the hydration of the C_3A must be retarded, and to this end gypsum is added during the grinding of the cement clinker (section 1.1). The gypsum combines with the C_3A to give a high-sulphate calcium sulphoaluminate, known as ettringite ($3CaO \cdot Al_2O_3 \cdot 3CaSO_4 \cdot 31H_2O$), and this formation of ettringite prevents the direct hydration of the C_3A and the resulting flash setting.

There is an 'optimum gypsum content' which imparts to the cement maximum strength and minimum shrinkage (Fig. 1.3), and this optimum depends on the alkali-oxides and the C_3A contents of the cement and on its fineness [1.3, 1.4]. On the other hand, the gypsum content must be limited because an excessive amount may cause cracking and deterioration in the set cement. This adverse effect is due to the formation of the ettringite which involves volume increase in the solids. When only a small amount of gypsum is added, the reaction takes place mainly when the paste or the concrete are plastic and the associated volume increase can be accommodated without causing any damage. When greater amounts are added, the formation of the ettringite, and the associated volume increase, take place also in the hardened cement and may cause, therefore, cracking and damage. Consequently, cement standards specify a maximum SO_3 content which depends on the type of cement considered and its C_3A content. In accordance with BS12, for

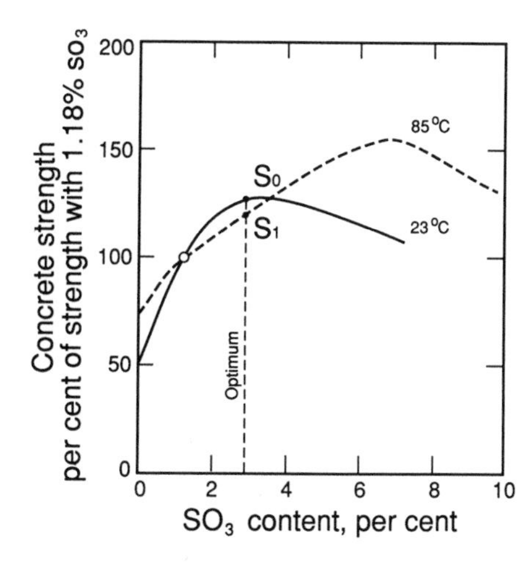

Fig. 1.4. Effect of temperature on optimum SO_3 content. (Adapted from Ref. 1.6).

example, this maximum is 2·5 and 3·5%, for low and high C_3A content cement, respectively (Table 1.2). Similar restrictions of the SO_3 content, but not exactly the same, are specified by the relevant ASTM Standard (Table 1.3) and, indeed, by all cement standards.

In cements with a C_3A content lower than 6%, the optimum SO_3 content may be as low as 2% for low alkali contents (i.e. below 0·5%) increasing to 3–4% as the alkali contents rise to 1%. In cements high in C_3A (i.e. more than 10%) the optimum SO_3 content is about 2·5–3% and 3·5–4% for low and high alkali contents, respectively [1.5]. It may be noted that the above-mentioned values are within the limitations imposed by the standards and, indeed, in the manufacture of Portland cements an attempt is made to add the gypsum in the amount which imparts to the cement the optimum content.

The optimum gypsum content is temperature-dependent and increases with an increase in the latter. Hence, the preceding optimum contents are valid only for conditions where hydration takes place under normal temperatures. This effect of temperature is demonstrated in Fig. 1.4, and it can be seen that, under the specific conditions considered, the optimum SO_3 content at 85°C significantly exceeded the maximum imposed by the standards, and reached some 7%. It follows that a cement with a SO_3 content which complies with the standards, would produce a lower strength in a concrete subjected to elevated temperatures than in otherwise the same concrete subjected to normal temperatures.

The effect of temperature on optimum SO_3 content is reflected in Fig. 1.4 by the difference $S_0 - S_1$, and may partly explain the adverse effect of elevated temperatures on concrete later-age strength. This adverse effect, however, is discussed in some detail further in the text (see section 6.6).

Table 1.2. Required Properties of Portland Cements in Accordance with British Standards

Property/component	*Type of cement (standard)*			
	OPC	*RHPC*	*LHPC*	*SRPC*
	BS 12, 1989		*BS 1370, 1979*	*BS 4027, 1980*
Chemical composition				
(BS 4550, Part 2, 1970)				
(1) Lime saturation factor (LST)	—	—	0·66—0·88	0·66—1·02
(2) Insoluble residue, max (%)	1·5	1·5	1·5	1·5
(3) Magnesia (MgO), max (%)	4·0	4·0	4·0	4·0
(4) Sulphuric anhydride (SO_3), max (%)	—	—	—	2·5
3·5% $<$ C_3A	3·5	3·5	—	—
3·5% $\geqslant$ C_3A	2·5	2·5	—	—
5·0% $<$ C_3A	—	—	3·0	—
5·0% $\geqslant$ C_3A	—	—	2·5	—
(5) Loss of ignition, max (%)	3·0	3·0	—	—
Temperate climate	—	—	3·0	3·0
Tropical climate	—	—	4·0	4·0
Mineralogical composition				
(1) Tricalcium—aluminate (C_3A), max (%)	—	—	—	3·5
Physical properties				
(BS 4550, Part 3, 1978)				
(1) Fineness, min (m^2/kg)	275	350	275	250
(2) Setting times, Vicat apparatus,				
Initial, min (min)	45	45	45	45
Final, max (h)	10	10	10	10

(3) Soundness, max (mm)	10	10	10	10
(4) Compressive strength (N/mm^2)				
(4.1) 100 mm concrete cubes at the age of				
2 days, min	—	15	—	—
3 days, min	15	—	5	10
28 days, min	34	38	19	27
28 days, max	52	—	—	—
(4.2) 70·7 mm mortar cubes (N/mm^2) at the age of				
2 days	—	25	—	—
3 days, min	25	—	10	20
28 days, min	47	52	28	39
28 days, max	67	—	—	—
(5) Heat of hydration, max (kJ/kg) at the age of				
7 days	—	—	250	—
28 days	—	—	290	—

Table 1.3. Required Properties of Portland Cements in Accordance with ASTM C150—89

Property/constituent	*Type and designation*				
	I *OPC*	*II* *Moderate*	*III* *RHPC*	*IV* *LHPC*	*V* *SRPC*
Chemical composition (ASTM C114)					
(1) Silicon dioxide, (SiO_2), min (%)	—	20·0	—	—	—
(2) Aluminium oxide (Al_2O_3), max (%)	—	6·0	—	—	—
(3) Ferric oxide (Fe_2O_3), max (%)	—	6·0	—	6·5	—
(4) Magnesium oxide (MgO), max (%)	6·0	6·0	6·0	6·0	6·0
(5) Sulphur trioxide (SO_3), max (%)					
C_3A content $\leqslant 8\%$	3·0	3·0	3·5	2·3	2·3
C_3A content $> 8\%$	3·5	—	4·5	—	—
(6) Loss on ignition, max (%)	3·0	3·0	3·0	2·5	3·0
(7) Insoluble residue, max (%)	0·75	0·75	0·75	0·75	0·75
Mineralogical composition					
(1) Tricalcium silicate (C_3S), max (%)	—	—	—	35	—
(2) Dicalcium silicate (C_2S), min (%)	—	—	—	40	—
(3) Tricalcium aluminate (C_3A), max (%)	—	8	15	7	5
(4) $C_4AF + 2(C_3A)$, max (%)	—	—	—	—	25
Physical properties					
(1) Specific surface area (Blaine), min (m^2/kg) (ASTM C204)	280	280	—	280	280
(2) Autoclave expansion, max (%) (ASTM C151)	0·80	0·80	0·80	0·80	0·80
(3) Time of setting, Vicat apparatus (ASTM C191)					
Initial set, min (not less than)	45	45	45	45	45
Final set, min (not more than)	375	375	375	375	375
(4) Compressive strength, min (MPa) at the age of					
1 day	—	—	12·4	—	—
3 days	12·4	10·3	24·1	—	8·3
7 days	19·3	17·2	—	6·9	15·2
28 days	—	—	—	17·2	20·7
(5) Heat of hydration, (optional) max (kJ/kg) at					
7 days	—	70	—	250	—
28 days	—	—	—	290	—
(6) Sulphate expansion at 14 days (ASTM C452), max (%) (optional)	—	—	—	—	0·040

1.3.2. Free Lime (CaO)

Lime makes up some 65% of the raw materials which are used to produce Portland cement. On clinkering, however, the lime combines with the other oxides of the raw materials to give the four major constituents of Portland cement discussed earlier. The presence of free (i.e. uncombined) lime in the cement may occur when the raw materials contain more lime than can combine with the acidic oxides SiO_2, Al_2O_3 and Fe_2O_3, or when the burning of the raw materials is not complete. Such incomplete burning may occur, for example, when the raw materials are not finely ground and intimately mixed. The presence of free lime in the cement may also result from an excessive content of phosphorous pentoxide (P_2O_5) in the raw materials [1.7]. Nevertheless, even under carefully controlled production, a small amount of free lime, usually less than 1%, remains in the clinker. Such a lime content, however, is not harmful.

The uncombined lime which remains in the cement is 'hard burnt', and as such is very slow to hydrate. Moreover, this lime is intercrystallised with other minerals and is, therefore, not readily accessible to water. Hence, the hydration of the free lime takes place after the cement has set. Since the hydration of lime to calcium hydroxide (slaked lime) involves a volume increase, the expansion of the latter may cause cracking and deterioration. Cements which exhibit such an expansion are said to be 'unsound' and the phenomenon is known as 'unsoundness due to lime'.

In view of the preceding discussion, it is clearly understandable that the free lime content of the cement must be limited. This limitation of the free lime content is usually imposed in the cement standards by specifying a minimum expansion of the set cement due to its exposure to curing conditions that cause the hydration of the free lime in a short time (Table 1.2). A relevant test, using the Le Chatelier apparatus, is described in BS 4550, Part 3, Section 3.7, 1978.

1.3.3. Magnesia (MgO)

The raw materials used for producing cement usually contain a small amount of magnesium carbonate ($MgCO_3$). Similarly to calcium carbonate, the $MgCO_3$ dissociates on burning to give magnesium oxide (magnesia) and carbon dioxide. The magnesia does not combine with the oxides of the raw materials and mostly crystallises to the mineral known as periclase. At the burning temperature of the cement, the magnesia is dead burnt and reacts with water, therefore, very slowly at ordinary temperatures. As the hydration of the magnesia (i.e. its conversion to $Mg(OH)_2$) involves volume increase, its presence in the cement in excessive amount may also cause unsoundness. Consequently, the magnesia content in the cement is limited to a few percent, i.e. to 4% in accordance with BS 12, 1989 (Table 1.2) or to 6% in accordance with ASTM C150 (Table 1.3).

1.3.4. Alkali Oxides (K_2O, Na_2O)

The alkali oxides are introduced into the cement through the raw materials, and their content usually varies from 0·5 to 1·3%.

The presence of the alkali in the cement becomes of practical importance when alkali-reactive aggregates are used in concrete production. Such aggregates contain a reactive form of silica or, much less frequently, a reactive form of carbonate, which combines with the alkali oxides of the cement. The reactions involved produce expansive forces which, in turn, may cause cracking and deterioration in the hardened concrete (see section 9.4). Generally speaking, this adverse effect may be avoided by using 'low-alkali' cements, i.e. cements in which the total alkali content, R_2O, calculated as equivalent to Na_2O, does not exceed 0·6%. The molar ratio Na_2O/K_2O equals 0·658. Hence, the Na_2O equivalent R_2O content is given by $R_2O = Na_2O + 0{\cdot}658\ K_2O$.

1.4. FINENESS OF THE CEMENT

Fineness of the cement is usually measured by its specific surface area, i.e. by the total surface area of all grains contained in a unit weight of the cement. Accordingly, the smaller the grain size, the greater the specific surface area, and vice versa.

The fineness of the cement affects its properties, and this effect manifests itself through its effect on the rate of hydration. The hydration of the cement is discussed later in the text (see section 2.3), but it may be realised that its rate increases with an increase in the fineness of the cement. The smaller the cement grains, the greater the surface area which is exposed to water and, consequently, the higher the rate of hydration.

It will also be shown later (section 6.2.2), that the rate of strength develop-

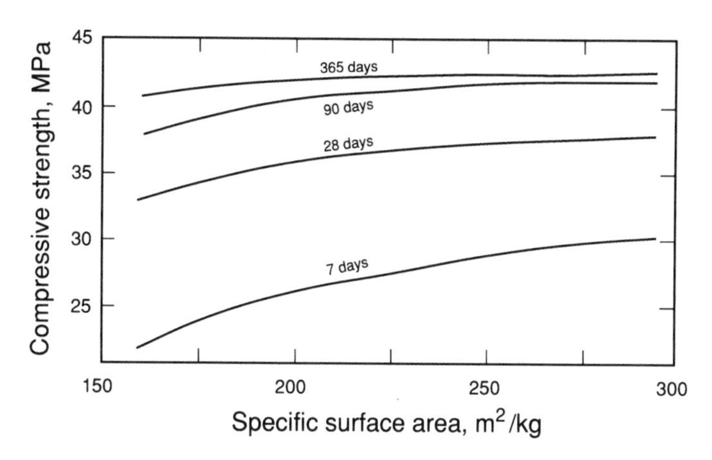

Fig. 1.5. Effect of specific surface area of Portland cement on strength development in concrete. (Adapted from Ref. 1.8.)

ment increases with the rate of hydration. It is expected, therefore, that the rate of strength development will increase with fineness as well. This is confirmed by the data of Fig. 1.5, which also indicate that the effect of fineness on strength is greatest at earlier ages, decreasing with time as the hydration proceeds. At later ages, as will be explained in some detail later in the text (see section 2.4), the cement grains become encapsulated in a dense layer of hydration products. This layer retards the diffusion of water, and thereby slows down hydration until, at some stage, it is stopped completely. The rate of hydration at this later stage is determined, therefore, by the rate of water diffusion, and the size of the cement grains becomes of secondary importance.

1.5. DIFFERENT TYPES OF PORTLAND CEMENT

As has already been mentioned, the properties of Portland cement, which is a heterogeneous material, are determined, qualitatively at least, by the properties of its major constituents. The properties of the latter are summarised in Table 1.1 and in Figs 1.1 and 1.2. It can be seen that the C_3A and C_4AF contribute only slightly to the strength of the cement, and that this is particularly true with respect to later-age strength which is mainly determined by the calcium silicates, i.e. by the Alite and the Belite. On the other hand, the C_3A increases the susceptibility of the cement to sulphate attack, its heat of hydration and probably its shrinkage as well. Accordingly, it may be

Table 1.4. Approximate Composition and Fineness Ranges for the Standard Types of Portland Cements[a]

ASTM type	*Tricalcium silicate (C_3S) (%)*	*Dicalcium silicate (C_2S) (%)*	*Tricalcium aluminate (C_3A) (%)*	*Tetracalcium aluminoferrite (C_4AF)[b] (%)*	*Air permeability specific surface area (m^2/kg)*
I	42–65	10–30	0–17	6–18	300–400
II	35–60	15–35	0–8	6–18	280–380
III	45–70	10–30	0–15	6–18	450–600
IV	20–30	50–55	3–6	8–15	280–320
V	40–60	15–40	0–5	10–18	290–350

[a]*Adapted from Ref. 1.9.*
[b]*C_4AF is actually a solid solution whose composition may vary considerably from one cement to another.*

concluded that it is desirable to increase the total calcium silicates content in the cement at the expense of the C_3A and C_4AF contents, and particularly at the expense of that of the C_3A. The reduction in the C_3A and C_4AF contents can be achieved by using raw materials poor in Al_2O_3 and Fe_2O_3. Such a reduction, however, is practical only to a limited extent because the presence of both Al_2O_3 and Fe_2O_3 in the raw materials lowers the clinkering temperature and is required, therefore, for economic and technical reasons. Consequently, the combined content of the calcium silicates is usually kept between 70 and 75% and that of the C_3A and the C_4AF between 15 and 20%. That is, in changing the cement composition, a variation in the C_3S content usually involves a corresponding variation in the opposite direction in the C_2S content. Similarly, a variation in the C_3A content usually involves an opposite variation in the C_4AF content (Table 1.4). Nevertheless, within these limitations, the composition of Portland cement can be changed to produce cements of different properties, i.e. to produce different types of Portland cement. In addition to 'ordinary' (sometimes 'normal') Portland cement, RHPC, LHPC, SRPC and white cements are produced.

1.5.1. Rapid-Hardening Cement (RHPC)

This type of cement, type III in accordance with ASTM C150, is characterised by a higher rate of strength development and, therefore, also by a higher early strength when compared with OPC (type I in accordance with ASTM C150). This difference in strength decreases with time and, as long as adequate curing

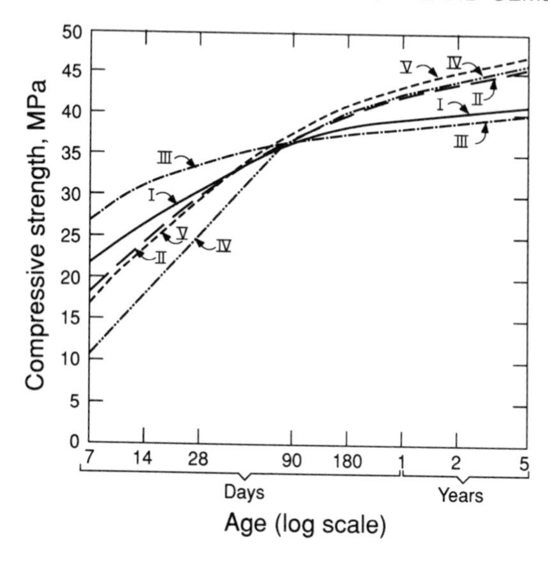

Fig. 1.6. Effect of type of cement on concrete strength development. (Adapted from Ref. 1.10).

is provided, both types may reach essentially the same level of strength (Fig. 1.6). This is usually the case when both cements are roughly of the same fineness, but when the RHPC is ground to a greater fineness, its strength may remain higher at later ages as well, and particularly when short curing periods are involved.

RHPC is produced by increasing the C_3S content at the expense of that of the C_2S (Table 1.4), and is usually ground to a greater fineness than the other types of cement. Accordingly, BS 12, 1989, for example, specifies a minimum specific surface area of 350 m^2/kg for RHPC as compared with 275 m^2/kg for OPC (Table 1.2).

The heat of hydration of C_3S is greater than that of C_2S. Hence, the heat of hydration of RHPC is greater than that of OPC. Moreover, the greater fineness, when this is the case, increases the rate of hydration of RHPC bringing about a corresponding increase in heat evolution (Fig. 1.7).

1.5.2. Low-Heat Cement (LHPC)

The heats of hydration of C_3S and C_3A are higher than those of the remaining constituents of the cement (Table 1.1). Accordingly, the heat of hydration of the cement can be lowered by reducing the contents of the C_3S and the C_3A (Table 1.4). It can be seen from Fig. 1.7 that, indeed, heat evolution in such a cement (type IV in accordance with ASTM C150) is lower than in all other types of Portland cement.

The heat of hydration of OPC varies from 420 to 500 J/g whereas, in

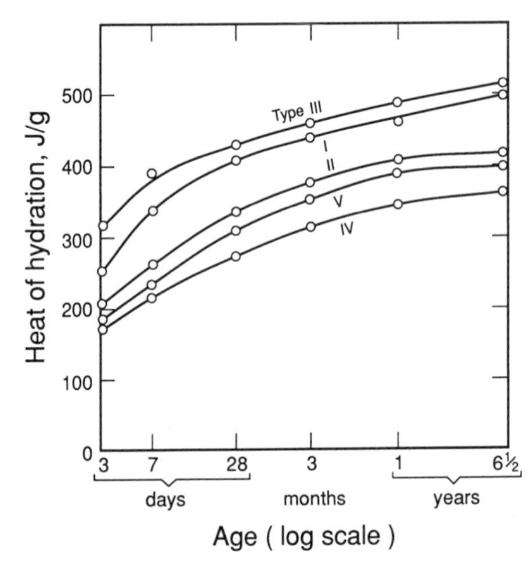

Fig. 1.7. Heat evolution in concrete made of different types of Portland cement. (Adapted from Ref. 1.11.)

accordance with ASTM C150 and BS 1370, 1979, the heat of hydration of LHPC should not exceed 250 J/g at the age of 7 days and 290 J/g at the age of 28 days (Tables 1.2 and 1.3).

In view of the comparatively low C_3S and C_3A contents in LHPC, it is to be expected that the rate of strength development in such a cement will be slow, and consequently its early age strength will be low. This is, of course, the case. On the other hand, the ultimate strength of LHPC may be even higher than that of ordinary or RHPC (Fig. 1.6).

The presence of C_3A makes the cement vulnerable to sulphate attack. Hence, the reduced C_3A content improves the sulphate-resisting properties of LHPC. In fact, as it will be seen in section 1.5.3, the properties of LHPC are similar to those of sulphate-resisting cement.

1.5.3. Sulphate Resisting Cement (SRPC)

Portland cement is vulnerable to sulphate attack and this vulnerability is mainly due to the presence of C_3A. The mechanism of sulphate attack is described later in the text (see section 9.3.1). This attack, however, involves volume increase and expansion which, in turn, may cause cracking and subsequently lead to severe deterioration. It can be seen from Fig. 1.8 that, indeed, sulphate expansion decreases with the decrease in the C_3A content and it is to be expected, therefore, that sulphate resistance of the cement will increase with a decrease in its C_3A content. This conclusion is widely recog-

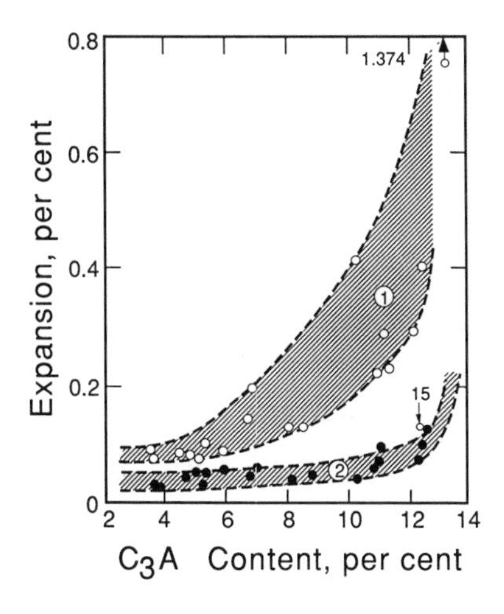

Fig. 1.8. Effect of C_3A content on potential sulphate expansion of Portland cement mortars after (1) 1 year, and (2) 1 month. (Adapted from Ref. 1.12.)

nised and utilised in the production of SRPC, i.e. in the production of cement type V in accordance with ASTM C150. Accordingly, BS 4027, 1980 limits the C_3A content in SRPC to 3·5% and ASTM C150-89 to 5% (Tables 1.2 and 1.3).

The presence of the ferrite phase (C_4AF) also makes the cement vulnerable to sulphate attack but its effect is much smaller than that of the C_3A. Consequently, whereas the C_3A content of SRPC is always limited, this is not necessarily the case with respect to the C_4AF. BS 4027, 1980, for example, does not impose any limits on the content of the latter. On the other hand, ASTM C150-89 specifies a maximum content which is related to that of the C_3A by the requirement that $C_4AF + 2(C_3A) \leqslant 25\%$ (Table 1.3).

The reduction in C_3A content reduces the heat of hydration of the cement (Fig. 1.7) and brings about slower rate of strength development and a lower early-age strength. Again, similar to LHPC, the later-age strength of a SRPC may sometimes be even higher than that of ordinary cement (Fig. 1.6). That is, SRPC exhibits similar properties to those of LHPC and, indeed, in certain applications the two are interchangeable. This observation is reflected, to some extent, in ASTM C150 which also includes a type II cement. This type is recommended for use when moderate sulphate resistance or, alternatively, moderate heat of hydration, are required.

1.5.4. White and Coloured Cements

The characteristic grey colour of Portland cement is due mainly to the presence of the ferrite phase. Accordingly, a white cement is produced by clinkering a mixture of raw materials low in iron oxide, i.e. usually lower than 0·5%. Consequently, the ferrite phase content of the resulting white cement is in the order of about 1% only. Generally, the strength of the white cement is lower than that of ordinary cement but, in many cases, it complies with the strength requirements specified in the standards for OPC.

Coloured cements are produced by grinding together a white clinker with a suitable pigment, or by blending the white cement with the desired pigment at a later stage. Both white and coloured cements are expensive to produce and their use is, therefore, limited to decorative and architectural applications in which only comparatively small quantities of the cement are required.

1.6. SUMMARY AND CONCLUDING REMARKS

Portland cement is an active hydraulic binder which is produced by clinkering a mixture of raw materials containing lime (CaO), silica (SiO_2), alumina (Al_2O_3) and iron oxide (Fe_2O_3), and grinding the resulting clinker with a few percent of gypsum. The cement produced in such a way is a heterogeneous material which is made of four 'major constituents', namely, Alite (essentially tricalcium silicate), Belite (essentially dicalcium silicate), an aluminate phase (essentially tricalcium aluminate) and a ferrite phase known as Celite (average composition approximately tetracalcium aluminoferrite). The combined total of the four major constituents is approximately 90%. The remaining 10% are collectively known as 'minor constituents', and include, in addition to gypsum (5%), free lime (1%), magnesia (2%) and the alkali oxides Na_2O and K_2O (1%).

Portland cement is a generic term for various types of cement which include, in addition to ordinary Portland cement (OPC), rapid-hardening cement (RHPC), low-heat cement (LHPC), sulphate-resisting cement (SRPC) and several others. These different types are produced by changing the composition of the cement and, sometimes, also by grinding the clinker to a different fineness.

REFERENCES

1.1. Bogue, R.H. & Lerch, W., Hydration of Portland cement compounds. *Ind. Engng Chem.*, **26**(8) (1934), 837–47.

1.2. Copeland, L.E. & Kantro, D.L., Kinetics of the hydration of Portland cement. In *Proc. Symp. Chem. Cement*, Washington, 1960, National Bureau of Standards Monograph No. 43, Washington, 1962, pp. 443–53.

1.3. Lerch, W., The influence of gypsum on the hydration properties of Portland cement pastes. *Proc. ASTM*, **46** (1946), 1252–92.

1.4. Meissner, H.S., *et al.*, The optimum gypsum content of Portland cement. *Bull. ASTM*, **169** (1950), 39–45.

1.5. Lea, F.M., *The Chemistry of Cement and Concrete* (2nd edn). Edward Arnold, London, UK, 1970, p. 308.

1.6. Verbeck, G. & Copeland, L.E., Some physical and chemical aspects of high pressure steam curing. In *Menzel Symposium on High Pressure Steam Curing* (ACI Spec. Publ. SP 32). ACI, Detroit, USA, 1972, pp. 1–13.

1.7. Gutt, W., Manufacture of Portland cement from phosphatic raw materials. In *Proc. Symp. Chem. Cement*, Tokyo, 1968, The Cement Association of Japan, Tokyo, pp. 93–105.

1.8. Price, W.H., Factors influencing concrete strength. *J. ACI*, **47**(2) (1951), 417–32.

1.9. ACI Committee 225, Guide to selection and use of hydraulic cements. (ACI 225R-85). In *ACI Manual of Concrete Practice* (Part 1). ACI, Detroit, MI, USA, 1990.

1.10. US Bureau of Reclamation, *Concrete Manual* (8th edn). Denver, CO, USA, 1975, p. 45.

1.11. Verbeck, G.J. & Foster, C.W., Long time study of cement performance in concrete. *The Heats of Hydration of the Cements*. Proc. ASTM **50** (1950) 1235–57.

1.12. Mahter, B., Field and laboratory studies of the sulphate resistance of concrete. In *Performance of Concrete*, ed. E.G. Swenson. University of Toronto Press, Toronto, Canada, 1968, pp. 66–76.

Chapter 2
Setting and Hardening

2.1. INTRODUCTION

Setting and hardening of cement can be described and discussed from three different points of view—phenomenological, chemical and structural. The phenomenological point of view, by definition, is concerned with the changes in the cement–water system (or the concrete) which are only perceptible to or evidenced by the senses. The chemical point of view is concerned with the chemical reactions involved and the nature and composition of the reactions products. Finally, the structural point of view is concerned with the structure of the set cement, and with the possible changes in this structure with time. Hence, the following discussion is presented accordingly. This discussion mainly considers the cement paste, i.e. a paste which is produced as a result of mixing cement with water only. Nevertheless, it is valid and applicable to mortar and concrete as well because, under normal conditions, the aggregate is inert in the cement–water system and its presence, therefore, does not affect the processes involved.

2.2. THE PHENOMENA

Mixing cement with water produces a plastic and workable mix, commonly referred to as a 'cement paste'. These properties of the mix remain unchanged

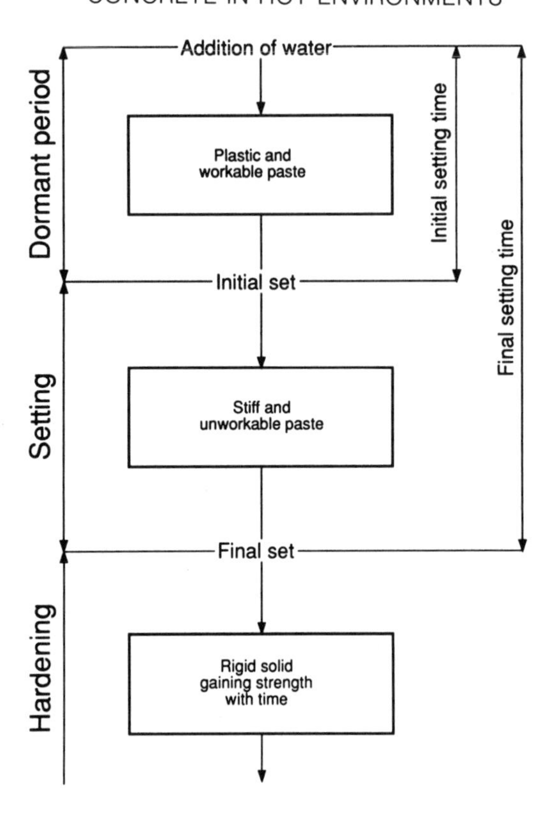

Fig. 2.1. Schematic description of setting and hardening of the cement paste. (Adapted from Ref. 2.1.)

for some time, a period which is known as the 'dormant period'. At a certain stage, however, the paste stiffens to such a degree that it loses its plasticity and becomes brittle and unworkable. This is known as the 'initial set', and the time required for the paste to reach this stage as the 'initial setting time'. A 'setting' period follows, during which the paste continues to stiffen until it becomes a rigid solid, i.e. 'final set' is reached. Similarly, the time required for the paste to reach final set is known as 'final setting time'. The resulting solid is known as the 'set cement' or the 'hardened cement paste'. The hardened paste continues to gain strength with time, a process which is known as 'hardening'. These stages of setting and hardening are schematically described in Fig. 2.1.

The initial and final setting times are of practical importance. The initial setting time determines the length of time in which cement mixes, including concrete, remain plastic and workable, and can be handled and used on the building site. Accordingly, a minimum of 45 min is specified in most standards for ordinary Portland cement (OPC) (BS 12, ASTM C150). On the other hand, a maximum of 10 h (BS 12) or 375 min (ASTM C150) is specified for the final setting time (see Tables 1.3 and 1.4). The need for such a maximum is required in order to allow the construction work to continue within a reasonable time after placing and finishing the concrete.

The setting times of the cement depend on its fineness and composition, and are determined, somewhat arbitrarily, from the resistance to penetration of the paste to a standard needle, using an apparatus known as the Vicat needle (BS 4550, Part 3; ASTM C191). In determining the setting times of concrete, in principle, essentially the same procedure is employed. The penetration resistance is determined, however, on a mortar sieved from the concrete through a 4·75 mm sieve, by a different apparatus sometimes known as the Proctor needle (ASTM C403).

Finally, setting times are affected by ambient temperature and are usually reduced with a rise in the latter. This specific effect of temperature on setting times is discussed later in the text (see section 2.6.1).

2.3. HYDRATION

In contact with water the cement hydrates (i.e. combines with water) to give a porous solid usually defined as a rigid gel (see section 2.4). Generally, chemical reactions may take place either by a through solution or by a topochemical mechanism. In the first case, the reactants dissolve and produce ions in solution. The ions then combine and the resulting products precipitate from the solution. In the second case, the reactions take place on the surface of the solid without its constituents going into solution. Hence, reference is made to topochemical or liquid–solid reactions. In the hydration of the cement both mechanisms are involved. It is usually accepted that the through-solution mechanism predominates in the early stages of the hydration, whereas the topochemical mechanism predominates during the later ones.

It was pointed out earlier that unhydrated cement is a heterogeneous material and it is to be expected, therefore, that its hydration products would vary in accordance with the specific reacting constituents. This is, of course, the case but, generally speaking, the hydration products are mainly calcium and aluminium hydrates and lime. In this respect the calcium silicate hydrates are, by far, the most important products. These hydrates are the hydration products of both the Alite and the Belite which make up some 70% of the cement. Hence, the set cement consists mainly of calcium silicate hydrates which, therefore, significantly determine its properties.

The calcium silicate hydrates are poorly crystallised, and produce a porous solid which is made of colloidal-size particles held together by cohesion forces and chemical bonds. Such a solid is referred to as a rigid gel and is further discussed in section 2.4.

The calcium silicate hydrates are sometimes assumed to have the average approximate composition of $3CaO \cdot 2SiO_2 \cdot 3H_2O$ ($C_3S_2H_3$). However, their exact composition and structure are not always clear and depend on several factors such as age, water to solid ratio and temperature. Consequently, in order to avoid implying any particular composition or structure, it is preferred to refer to the hydrates in question by the non-specific term of 'calcium silicate hydrates'. Similarly, the general term CSH is used to denote the composition of the calcium hydrates of the cement.

In addition to the calcium silicate hydrates, the hydration of both the Alite and the Belite produces a considerable quantity of lime (calcium hydroxide), i.e. some 40% and 18% of the total hydration products of the Alite and the Belite, respectively. The presence of calcium hydroxide in such a large quantity in the set cement has very important practical implications. It makes the cement paste, as well as the concrete, highly alkaline (i.e. the pH of the pore water exceeds 12·5), and explains, in turn, why Portland cement concrete is very vulnerable to acid attack, and why concrete, unless externally protected, is unsuitable for use in an acidic environment. It is much more important, in this respect, that the corrosion of steel is inhibited once the pH of its immediate environment exceeds, say, 9. That is, unless the $Ca(OH)_2$ is carbonated, concrete provides the steel with adequate protection against corrosion. This protective effect of the alkaline surroundings is, of course, very important with respect to the durability of reinforced concrete structures, and is further discussed in Chapter 10.

The hydration of the cement results in heat evolution usually referred to as the heat of hydration. The heat of hydration of OPC varies, depending on its mineralogical composition, from 420 to 500 J/g. The relation between the mineralogical composition and heat of hydration, and the utilisation of this relation to produce low-heat Portland cement, were discussed earlier in section 1.5.2.

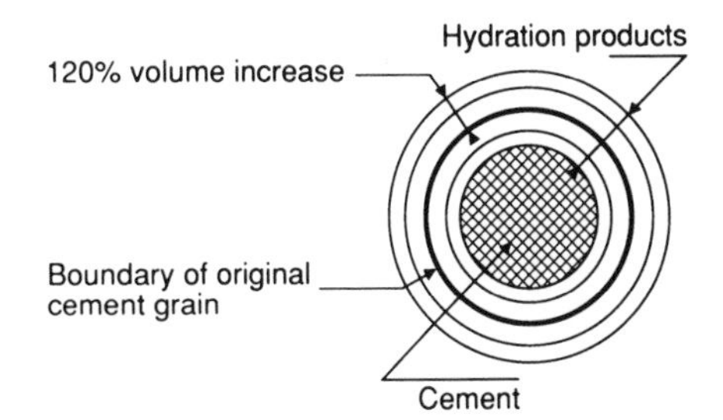

Fig. 2.2. Schematic description of the hydration of a cement grain.

2.4. FORMATION OF STRUCTURE

It was pointed out in section 2.3 that at a later stage the hydration reactions are essentially of a topochemical nature and as such take place mostly on the surface of the cement. Consequently, the hydration products are deposited on the surface and form a dense layer which encapsulates the cement grains (Fig. 2.2). As the hydration proceeds, the thickness of the layer increases, and the rate of hydration decreases because it is conditional, to a great extent, on the diffusion of water through the layer. That is, the greater the thickness of the layer, the slower the hydration rate explaining, in turn, the nature of the observed decline in the rate of hydration with time (Fig. 2.3). Moreover, it is to be expected that, after some time, a thickness is reached which hinders further diffusion of water, and thereby causes the hydration to cease even in the presence of a sufficient amount of water. This limiting thickness is about 10 μm, implying that unhydrated cores will always remain inside cement grains having a diameter greater than, say, 20 μm. This conclusion explains, partly at least, why the cement standards impose restrictions on the coarseness of the cement, usually by specifying a minimum specific surface area (see Tables 1.3 and 1.4). Consequently, the size of the cement grains in OPC varies from 5 to 55 μm.

Structure formation in the hydrating cement paste is schematically described in Fig. 2.4. The total volume of the hydration products is some 2·2 times greater than the volume of the unhydrated cement (Fig. 2.2) and, consequently, the spacing between the cement grains decreases as the hydration

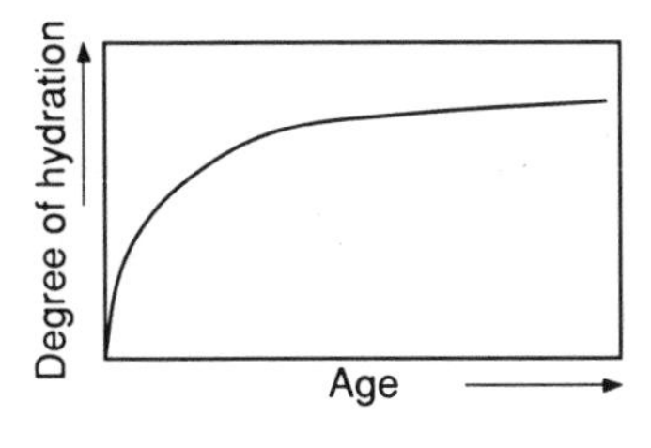

Fig. 2.3. Schematic description of the relation between the degree of hydration and time.

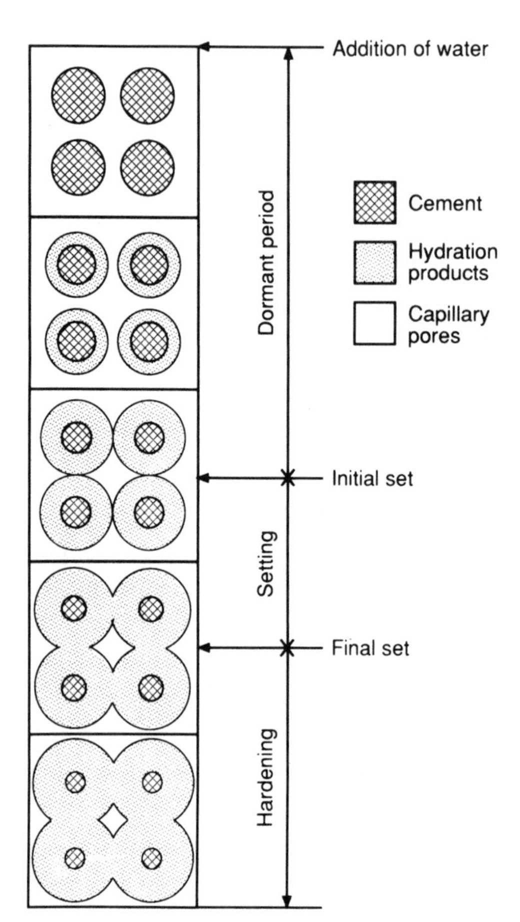

Fig. 2.4. Schematic description of structure formation in a cement paste.

proceeds. Nevertheless, for some time, the grains remain separated by a layer of water and the paste retains its plasticity and workability. This is the dormant period which was previously discussed (see section 2.2).

As the hydration proceeds the spacing between the cement grains further decreases, and at a certain stage friction between the hydrating grains is increased to such an extent that the paste becomes brittle and unworkable, i.e. 'initial set' is reached. On further hydration, bonds begin to form at the contact points of the hydrating grains, and bring about continuity in the structure of the cement paste. Consequently, the paste gradually stiffens and subsequently becomes a porous solid, i.e. 'final set' is reached. The resulting solid is characterised by a continuous pore system usually known as 'capillary porosity'. If water is available, the hydration continues and the capillary porosity decreases due to the formation of additional hydration products. It is to be expected that this decrease in porosity will result in a corresponding

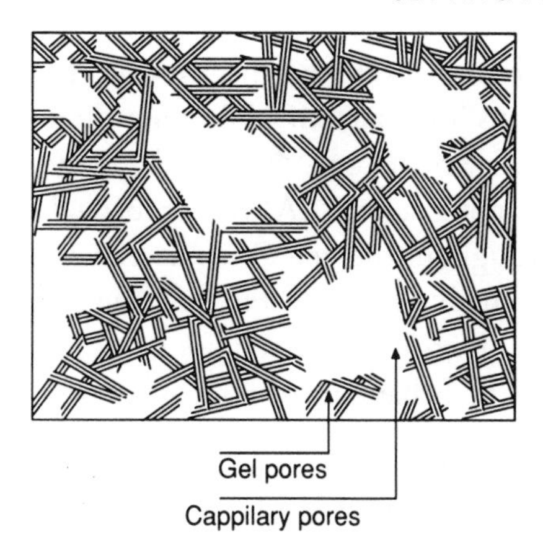

Fig. 2.5. Schematic description of the structure of the cement gel. (Adapted from Ref. 2.2.)

increase in the paste strength. This is, of course, the case, and this important aspect of the porosity–strength relationship is further discussed in section 6.2.

It was mentioned earlier (section 2.3), that the hydration products consist mainly of calcium silicate hydrates which produce a porous solid usually referred to as a rigid gel. A gel is comprised of solid particles of colloidal size, and its strength is determined, therefore, by the cohesion forces operating between the particles. Such a gel, however, is unstable and disintegrates on the adsorption of water, whereas the set cement is very stable in water. This latter characteristic of the set cement is attributed to chemical bonds which are formed at some contact points of the gel particles, and thereby impart to the gel its rigidity and stability in water. Hence, the reference to a 'rigid gel'.

The size of the gel particles is very small, indeed, and imparts to the gel a very great specific surface area which, when measured with water vapour, is of the order of 200 000 m^2/kg. The cohesion forces are surface properties and, as such, increase with the decrease in the particles size or, alternatively with the increase in their specific surface area. Accordingly, the mechanical strength of the set cement is attributable, partly at least, to the very great specific surface area of the cement gel.

The cement gel has a characteristic porosity of 28% with the size of the gel pores varying between 20 and 40 Å. The capillary pores mentioned earlier, which are the remains of the original water-filled spaces that have not become filled with hydration products, are much bigger. It can be realised that the volume of the capillary pores varies and depends, in the first instance, on the original water to cement (W/C) ratio and subsequently on the degree of hydration.

A schematic description of the structure of the cement gel is presented in Fig. 2.5, in which the gel particles are represented by two or three parallel lines to indicate the laminar nature of their structure. On the macro-scale, not shown in Fig. 2.5, unhydrated cement grains and calcium hydroxide (lime) crystals are detectable embedded in the cement gel. Air voids, either introduced intentionally by using air-entraining agents (AEA), or brought about by entrapped air, are also present throughout the gel. Of course, due to the porous nature of its structure, water is usually present in the set cement in an amount which varies in accordance with environmental conditions. Water plays a very important role in determining the behaviour of the paste, and is sometimes classified as follows [2.3]:

(1) Water which is combined in the hydration products and, as such, constitutes part of the solid. Such water has been referred to as 'chemically bound water', 'combined water' or 'non-evaporable water'. This type of water is used, sometimes, to determine quantitatively the degree of hydration.
(2) Water which is present in the gel pores and is known, accordingly, as 'gel water'. Due to the very small size of the gel pores, most of the gel water is held by surface forces and, accordingly, is considered as physically adsorbed water. As the mobility of this type of water is restricted by surface forces, such water is not chemically active.
(3) Water which is present in the bigger pores beyond the range of the surface forces of the solids of the paste. Such 'free' water is usually referred to as 'capillary water'.

2.5. EFFECT OF TEMPERATURE ON THE HYDRATION PROCESS

2.5.1. Effect on Rate of Hydration

The rate of chemical reactions, in general, increases with a rise in temperature, provided there is a continuous and uninterrupted supply of the reactants. This effect of temperature usually obeys the following empirical equation which is known as the Arrhenius equation:

$$\frac{d(\ln k)}{dT} = A/RT^2 \qquad (2.1)$$

in which k is the specific reaction velocity, T is the absolute temperature, A is a constant usually referred to as the energy of activation, and R is the gas law constant, i.e. $R = 8{\cdot}314\,J/mol°C$.

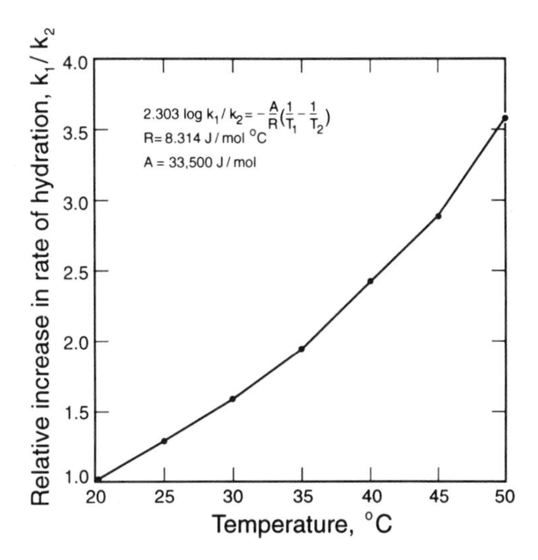

Fig. 2.6. Effect of temperature on the hydration rate of Portland cement in accordance with the Arrhenius equation.

It can be shown that, based on the former equation, the ratio between the rates of hydration k_1/k_2 at the temperatures T_1 and T_2, respectively, is given by the following equation:

$$2{\cdot}303 \log (k_1/k_2) = -\frac{A}{R}\left(\frac{1}{T_1} - \frac{1}{T_2}\right) \tag{2.2}$$

In the temperature range above 20°C, the energy of activation for Portland cement may be assumed to equal 33 500 J/mol [2.4]. Solving the equation accordingly (Fig. 2.6), it follows that the rise in the hydration temperature from $T_1 = 20°C$ to $T_2 = 30$, 40 and 50°C, will increase the hydration rate by factors of 1·57, 2·41, and 3·59, respectively. That is, the accelerating effect of temperature on the hydration rate of Portland cement is very significant indeed.

This expected accelerating effect of temperature is experienced, of course, in everyday practice and is supported by a considerable body of experimental data. It is clearly demonstrated, for example, in Fig. 2.7 in which the degree of hydration is expressed by the amount of the chemically bound water. Indeed, this accelerating effect of temperature is well known and recognised, and is widely utilised to accelerate strength development in concrete.

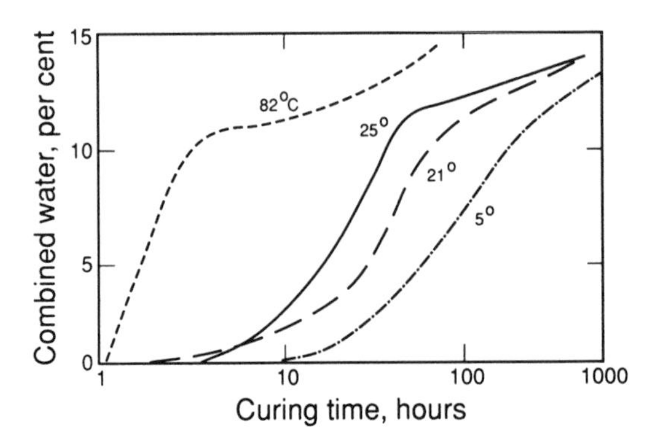

Fig. 2.7. Effect of temperature on the rate of hydration. (Adapted from Ref. 2.5.)

2.5.2. Effect on Ultimate Degree of Hydration

The effect of temperature on ultimate degree of hydration is not always clear. It was explained earlier (see section 2.4), that the ultimate degree of hydration is determined by the limiting thickness of the layer of the hydration products which is formed around the hydrating cement grains. The limiting thickness, as such, depends on the density of the gel layer, and the thickness of the latter and the associated ultimate degree of hydration, are expected to decrease with the increase of the gel density, and vice versa. Assuming, however, that gel density is not affected significantly by temperature, the ultimate degree of hydration is expected not to be affected by the temperature as well. This is supported by the data of Fig. 2.7 which indicate that essentially the same degree of hydration is reached in cement pastes regardless of the curing temperature. On the other hand, the data of Fig. 2.8 suggest that the ultimate degree of hydration increases with temperature while other data indicate the opposite, i.e. that the ultimate degree of hydration decreases [2.7]. It may be

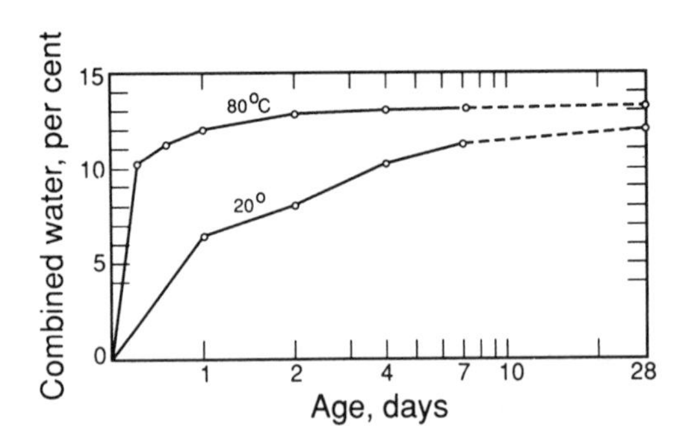

Fig. 2.8. Effect of temperature on the degree of hydration. (Adapted from Ref. 2.6.)

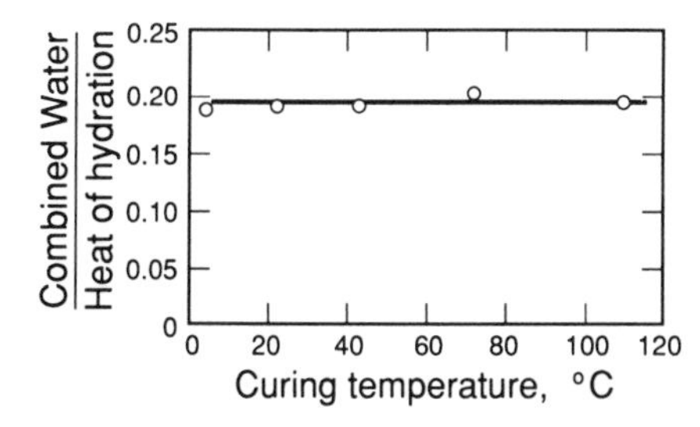

Fig. 2.9. Effect of curing temperature on the ratio of combined water to heat of hydration. (Adapted from Ref. 2.9.)

argued that, in the tests considered, the ultimate degree of hydration was not reached, thereby explaining, partly at least, the somewhat contradictory nature of the data in question. In any case, the temperature effect on ultimate degree of hydration is apparently small and of limited practical importance.

2.5.3. Effect on Nature of the Hydration Products

It is generally accepted that, in the temperature range up to 100°C, although the morphology and microstructure of the hydration products are somewhat affected, the stoichiometry of the hydration remains virtually the same and the hydration products do not differ essentially from those which are formed at moderate temperatures [2.8]. The similarity in the composition of the hydration products, regardless of curing temperature is somewhat supported by the data of Fig. 2.9. As both combined water and heat of hydration measure the degree of hydration, the observation that the ratio between the two remains constant implies that, at least as far as the chemically bound water is concerned, no change in composition is brought about by the change in curing temperature.

Some other data, which relate to C_2S and C_3S pastes indicate, however, that the composition of the hydration products is actually affected by curing temperatures, and in such pastes the CaO to SiO_2 ratio was found to increase, and the water to SiO_2 ratio to decrease, with the increase in temperature in the range 25–100°C [2.10]. In yet another study, however, such an increase was observed only in the temperature range of 25° to 65°C, but the trend was reversed in the lower range of 4–25°C [2.11]. In the latter study it was also found that the polysilicate content in the hydrated C_3S increased with time and the increase in temperature in the range 4–65°C.

It is not clear to what extent, if any, the preceding effects of temperature affect the performance and the mechanical properties of the set cement. In this context it should be pointed out that the latter properties are much more dependent on the structure of the set cement rather than on the exact

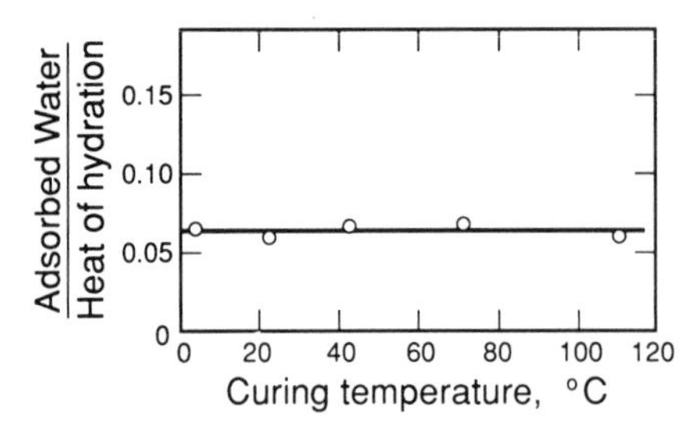

Fig. 2.10. Effect of curing temperature on the ratio of adsorbed water to heat of hydration. (Adapted from Ref. 2.9.)

composition of the hydration products. This is true only when no chemical corrosion is involved. Otherwise, the composition of the hydration products becomes very significant.

2.5.4. Effect on Structure of the Cement Gel

Temperature, through its accelerating effect on the rate of hydration (section 2.5.1), accelerates the formation of the gel structure. Temperature, however, also affects the nature of the structure as such and, in particular, the nature of its pore system. This effect is of practical importance because the mechanical properties of concrete, as well as its durability, are very much dependent on the physical characteristics of the gel structure.

Figure 2.10 presents data on the effect of temperature on the specific surface area of the cement gel. The ratio of adsorbed water to heat of hydration is equivalent to the ratio of the gel surface area to its content in the paste, i.e. it measures the gel specific surface area. The latter property of the gel remaining constant, implies also that the size of the gel particles is not affected by temperature. The strength of a rigid gel, such as the cement gel, depends, to a large extent, on the size of its particles. The size of the particles remaining the same implies that whatever is the effect of temperature on the strength of cement paste and concrete, this effect cannot be attributed to changes in the specific surface area of the cement gel. This aspect of strength is further discussed later in the text (see section 6.5).

In discussing the structure of the set cement (section 2.4), it was explained that porosity decreases as the hydration proceeds. Hence, as the rate of hydration is accelerated with temperature, the corresponding decrease in porosity is similarly accelerated. Consequently, at a certain age, the porosity of a paste cured at a lower temperature will be greater than the porosity of otherwise the same paste, cured at a higher temperature. It can be seen from Fig. 2.11 that this is, indeed, the case. As the hydration proceeds, however, this

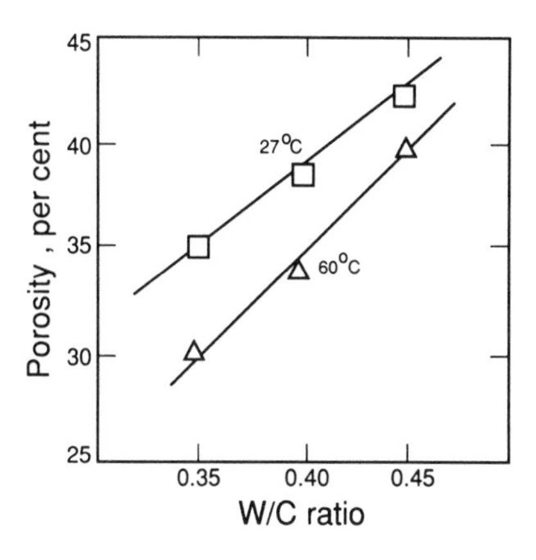

Fig. 2.11. Effect of W/C ratio and temperature on total porosity of a cement paste at 28 days. (Adapted from Ref. 2.12.)

effect of temperature on porosity becomes less evident because the effect of temperature on the ultimate degree of hydration is small (section 2.5.2).

On the other hand, temperature affects the nature of the pore-size distribution in the set cement, and a higher temperature is usually associated with a coarser system. This effect of temperature is demonstrated in Fig. 2.12 and was also observed by others [2.13, 2.14]. It can be seen that, although total porosity was lower in the paste which was cured at 60°C, the volume of pores with a radius greater than 750 Å was greater at the higher temperature. This is a very important observation because permeability of cement pastes is mostly determined by the volume of the larger pores rather than by total porosity (section 9.2). Moreover, the coarser nature of the pore system may also partly explain the adverse effect of temperature on later-age strength (section 6.5).

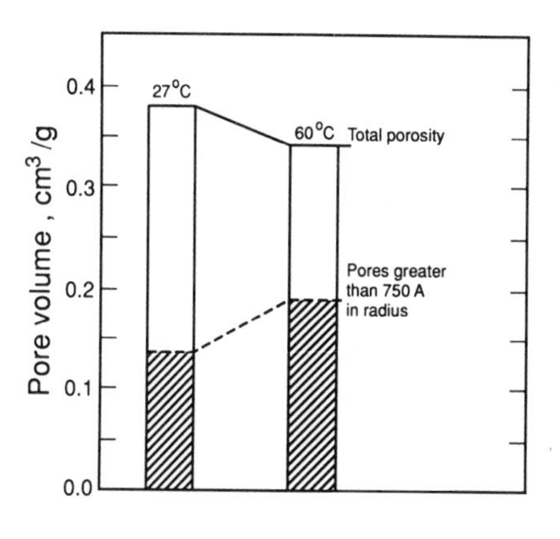

Fig. 2.12. Effect of temperature on total porosity and volume of pores having a radius greater than 750 Å. (Cement paste at 28 days, W/C ratio = 0·40.) (Adapted from the data in Ref. 2.12.)

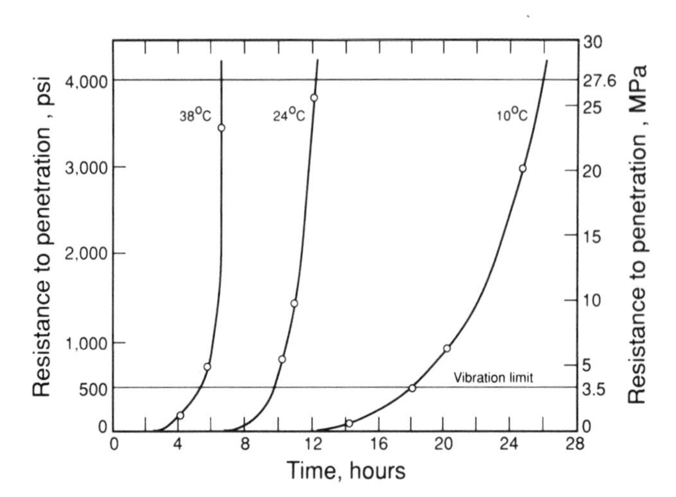

Fig. 2.13. Effect of temperature on setting of concrete (ASTM C403) (1 psi = 6·9 kPa). (Adapted from Ref. 2.15.)

2.6. EFFECT OF TEMPERATURE—PRACTICAL IMPLICATIONS

The accelerating effect of temperature on the rate of hydration manifests itself in three practical implications which are particularly relevant to concreting under hot-weather conditions. These include the reducing effect of temperature on setting times, its accelerating effect on the rate of stiffening (i.e. slump loss) and its increasing effect on the rate of temperature rise inside the concrete, and particularly inside mass concrete.

2.6.1. Effect on Setting Times

As a result of the accelerated hydration, initial and final setting times are both reduced with the rise in temperature. This effect of temperature is demonstrated, for example, in Fig. 2.13 in which the setting times are expressed by the resistance of the sieved concrete to penetration in accordance with ASTM C403 (see section 2.2). This effect of temperature is well recognised [2.16–2.18], and is more pronounced in the lower, than in the higher temperature range. It can be seen (Fig. 2.13) that a 14°C rise in temperature from 10 to 24°C reduced the initial setting time (i.e. vibration limit) by 8 h (i.e. from 18 to 10 h) while the same rise in temperature from 24 to 38°C reduced the latter by 5 h only (i.e. from 10 to 5 h).

2.6.2. Effect on Rate of Stiffening

The increased rate of hydration with temperature implies that the cement combines with water at a higher rate. The amount of free water in the mix is, consequently, reduced, bringing about the stiffening of the mix at a correspondingly higher rate. Moreover, the rate of stiffening is further increased by the more intensive drying of the mix with the rise in ambient temperature, particularly in dry environments. This effect of temperature on the rate of stiffening is, of course, well known and generally recognised, and is referred to in concrete technology as 'slump loss'. An accelerated slump loss is, of course, undesirable because it reduces the length of time during which the fresh concrete remains workable and can be handled properly at the building site. In fact, this phenomenon of slump loss constitutes one of the major problems of hot-weather concreting and is, therefore, discussed in some detail in section 4.4. Generally, however, in order to overcome the practical problems associated with the accelerated slump loss, one or more of the following means are employed:

(1) using a wetter mix, i.e. a mix of a higher slump, either by increasing the amount of mixing water or by the use of water-reducing admixtures;
(2) lowering concrete temperature by using cold mixing water or by substituting ice for part (up to 75%) of the mixing water;
(3) retempering, i.e. adding water or superplasticisers, or both, to the mix in order to restore the initial consistency of the concrete; and
(4) concreting during the cooler parts of the day, i.e. during the evening or at night.

2.6.3. Effect on Rise of Temperature

Concrete is a poor heat conductor, and the rate of heat evolution due to the hydration of the cement is, therefore, much greater than the rate of heat dissipation and, consequently, the temperature inside the concrete rises. With time, however, the inner concrete cools off and contracts, but this contraction is restrained to a greater or lesser extent. Restrained contraction results in tensile stresses, and this restraint may cause cracking if, and when, the tensile strength of the concrete at the time considered is lower than the induced stresses. The mode of restraint may be different, and in this respect reference is made to external and internal restraints. An external restraint takes place, for

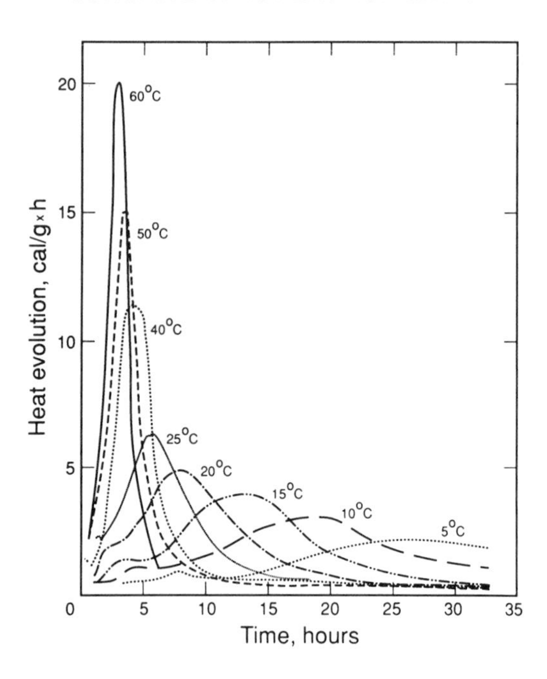

Fig. 2.14. Effect of temperature on heat evolution in the hydration of C_3S (1 cal = 4·2 J). (Adapted from Ref. 2.17.)

example, when new concrete is placed on top of an older one (e.g. a wall on a continuous foundation), and no separation is provided between the two. The internal restraint occurs always, and particularly in semi-mass or mass concrete, because the temperature of the outer layers of the concrete is close to the ambient temperature, whereas that of the internal core is always higher, and sometimes much higher. Hence, the thermal contraction of the internal core is restrained by the outer layers, and experience has shown that when the temperature difference between the inner and outer concrete exceeds, say, 20°C, cracking is liable to occur. It is implied, therefore, that in order to eliminate such cracking the rise in concrete temperature must be controlled accordingly. To this end several means are available, but these means are outside the scope of the present discussion.

It can be realised that this problem of thermal cracking is further aggravated by the accelerating effect of temperature on the rate of hydration. This effect results in a higher rate of heat evolution which, in turn, brings about a higher rise in concrete temperature. The increased rate of heat evolution with temperature in a C_3S paste is demonstrated in Fig. 2.14, and the increased rise in concrete temperature in Fig. 2.15.

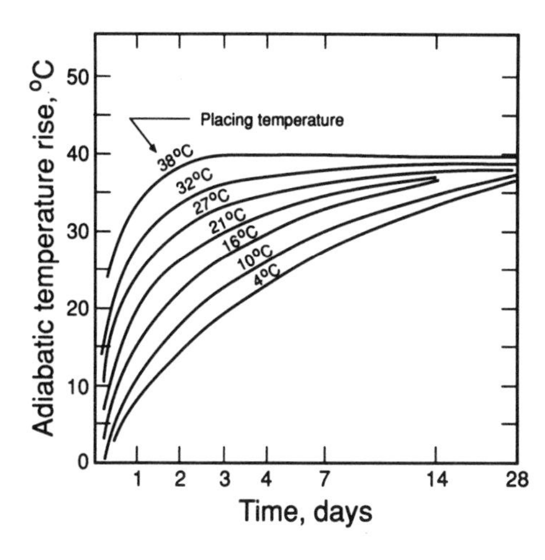

Fig. 2.15. Effect of placing temperature on temperature rise in mass concrete containing 223 kg/m^3 of type I cement. (Adapted from Ref. 2.18.)

2.7. SUMMARY AND CONCLUDING REMARKS

Mixing cement with water produces a plastic and workable mix known as a cement paste. The properties of the mix remain unchanged for some time, but at a certain stage it stiffens and becomes brittle and unworkable. This stage is known as initial setting. The setting period follows, and the paste continues to stiffen until it becomes a rigid solid, i.e. final setting is reached. The resulting solid continues to harden and gain strength with time, a process which is known as hardening.

Setting and hardening are brought about by the hydration of the cement. The hydration products are mainly hydrates of calcium silicates and lime, with the remaining ones being aluminates and ferrites. The hardened cement paste is a heterogeneous solid consisting of an apparently amorphous mass containing, mainly crystals of calcium hydroxide, unhydrated cement grains and voids containing either water or air, or both. The amorphous mass is a rigid gel made of colloid-size particles of calcium silicate hydrates, and has a characteristic fine porosity of 28% and a very large specific surface area. Much bigger pores, which are the remains of the original water-filled spaces that have not become filled with the hydration products, are also present in the gel and are known as capillary pores. The volume of the capillary pores decreases as the hydration proceeds because the volume of the hydration products is some 2·2 times greater than the volume of the reacting anhydrous cement. The decrease in porosity brings about a corresponding increase in strength.

The rate of hydration increases with temperature. Consequently, the rate of concrete stiffening (i.e. slump loss) is accelerated, its initial and final setting

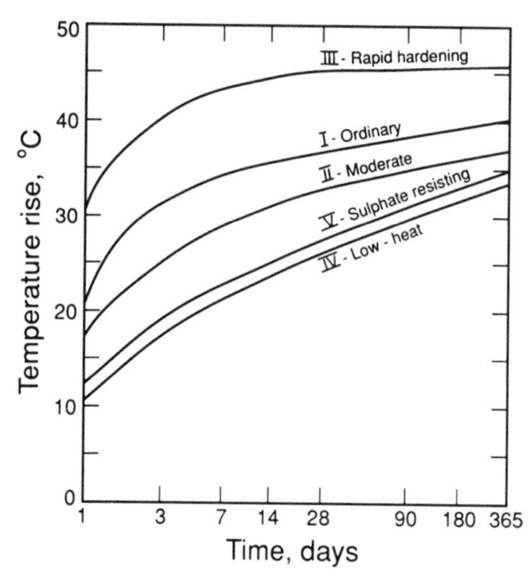

Fig. 2.16. Temperature rise in mass concrete made with 223 kg/m^3 cement of different types. (Adapted from Ref. 2.19.)

times are reduced, and the rise in concrete temperature is increased. Accordingly, it may be concluded that in hot weather conditions, the use of low-heat cement is to be preferred and the use of rapid-hardening cement must be avoided. This conclusion is clearly evident from Fig. 2.16, which indicates that the temperature inside a concrete made with rapid-hardening cement (type III) may be some 20°C higher than that inside a concrete made with low-heat cement (type IV).

The heat of hydration of blended cements, whether they are made of granulated blast-furnace slag, fly-ash or pozzolan, is lower than the heat of OPC. This property of blended cements is discussed in some detail in Chapter 3 and, indeed, the temperature rise in concrete made of such cements is lower than the rise in temperature in concrete made with OPC. Hence, from this point of view, the use of blended cements may be considered desirable in hot-weather conditions.

REFERENCES

2.1. Soroka, I., *Portland Cement Paste and Concrete*. The Macmillan Press Ltd, London, UK, 1979, p. 28.

2.2. Powers, T.C., Physical properties of cement paste. In *Proc. Symp. Chem. Cement*, Washington, 1960, National Bureau of Standards Monograph No. 43, Washington, 1962, pp. 577–613.

2.3. Powers, T.C. & Brownyard, T.L., Studies on the physical properties of

hardened Portland cement paste. Portland Cement Association Research Department Bulletin, No. 22, Chicago, MI, USA, 1948.
2.4. Hansen, P.F. & Pedersen, E.J., Curing of Concrete Structure. Report prepared for CEB—General Task Group No. 20, Danish Concrete and Structural Research Institute, Dec. 1984.
2.5. Taplin, J.H., The temperature dependence of the hydration rate of Portland cement paste. *Aus. J. Appl. Sci.*, **13**(2) (1962), 164–71.
2.6. Odler, I. & Gebauer, J., Cement hydration in heat treatment. *Zement–Kalk–Gips*, **55**(6) (1966), 276–81 (in German).
2.7. Idorn, G.M., Hydration of Portland cement paste at high temperatures under atmospheric pressure. In *Proc. Sump. Chem. Cement*, Tokyo, 1968, The Cement Association of Japan, Tokyo, pp. 411–35.
2.8. Taylor, H.F.W. & Roy, D.M., Structure and composition of hydrates. In *Proc. Symp. Chem. Cement*, Paris, 1980, Editions Septima, Paris, pp. II-2/1–2/13.
2.9. Verbeck, G.J. & Helmuth, R.H., Structure and physical properties of cement paste. In *Proc. Symp. Chem. Cement*, Tokyo, The Cement Association of Japan, Tokyo, pp. 1–37.
2.10. Odler, I. & Skalny, J., Pore structure of hydrated calcium silicates. *J. Colloid. Interface Sci.*, **40**(2) (1972), 199–205.
2.11. Bentur, A., Berger, R.L., Kung, J.H., Milestone, N.B. & Young, J.F., Structural properties of calcium silicate pastes: II, Effect of curing temperature. *J. Am. Ceramic Soc.*, **62**(7) (1977), 362–6.
2.12. Goto, S. & Roy, D.M., The effect of *W/C* ratio and curing temperature on the permeability of hardened cement paste. *Cement Concrete Res.*, **11**(4) (1981), 575–9.
2.13. Young, J.F., Berger, R.L. & Bentur, A., Shrinkage of tricalcium silicate pastes: Superposition of several mechanisms. *Il Cemento*, **75**(3) (1978), 391–8.
2.14. Kayyali, O.A., Effect of hot environment on the strength and porosity of Portland cement paste. *Durability of Building Materials*, **4**(2) (1986) 113–26.
2.15. Tuthill, L.H. & Cordon, W.A., Properties and uses of initially retarded concrete. *Proc. ACI*, **52**(3) (1955), 273–86.
2.16. Tuthill, L.H., Adams, R.F. & Hemme, J.M., Jr, Observation in testing and the use of water reducing retarders. In *Effect of Water Reducing Admixtures and Set Retarding Admixtures on Properties of Concrete.* (ASTM Spec. Tech. Publ. 266) Philadelphia, PA, USA, 1960.
2.17. Courtault, B. & Longuet, P., Flux adaptable calorimeter for studying heterogeneous solid–liquid reactions—Application to cement chemistry. *IVe Journees Nationales de Calorimetrier*, (1982), pp. 2/41–2/48 (in French).
2.18. ACI Committee 207, Effect of restraint, volume change, and reinforcement on cracking of massive concrete (ACI 207.2R-73) (Reaffirmed 1986). In *ACI Manual of Concrete Practice* (Part 1). ACI, Detroit, MI, USA, 1986.
2.19. ACI Committee 207, Mass Concrete (ACI 207.1R-87). In *ACI Manual of Concrete Practice* (Part 1). ACI, Detroit, MI, USA, 1990.

Chapter 3
Mineral Admixtures and Blended Cements

3.1. MINERAL ADMIXTURES

Admixtures are, by definition, 'a material other than water, aggregates, hydraulic cement and fibre reinforcement used as an ingredient of concrete or mortar and added to the batch immediately before or during mixing' (ASTM C125). Such a definition satisfies a wide range of materials, but a comprehensive discussion of all types involved is not attempted here. Accordingly, the following presentation is limited to the so-called 'mineral admixtures' whereas another group of admixtures, known as 'chemical admixtures' is discussed in section 4.3.2. The preceding reference to mineral admixtures is not always accepted and the term 'additions', rather than admixtures, has been suggested [3.1]. Moreover, this term of mineral additions was defined to include materials which are blended or interground with Portland cement, in quantities exceeding 5% by weight of the cement, and not only those which are added directly to the concrete before or during mixing. On the other hand, the term 'addition' was defined as 'a material that is interground or blended in limited amounts into hydraulic cement as a "processing addition" to aid manufacturing and handling of the cement, or as a "functional addition" to modify the use properties of the finished product' (ASTM C219). That is, the latter is quite a different definition which covers a different type of materials. Hence, in order to avoid possible misunderstanding, the term 'mineral admixtures', as defined by ASTM C125-88, is used hereafter.

Generally, mineral admixtures are finely divided solids which are added to the concrete mix in comparatively large amounts (i.e. exceeding 15% by weight of the cement) mainly in order to improve the workability of the fresh concrete and its durability, and sometimes also its strength, in the hardened state. It will be seen later (section 3.2) that these materials are also used as partial replacement of Portland cement in the production of 'blended cements'.

Mineral admixtures may be subdivided into low-activity, pozzolanic and cementitious admixtures.

3.1.1. Low-Activity Admixtures

This type of admixture, sometimes referred to as 'inert fillers', hardly reacts with water or cement and its effect is, therefore, essentially of a physical nature. Finely ground limestone or dolomite, for example, constitute such admixtures, and their use may be beneficial in improving the workability and the cohesiveness of concrete mixes which are deficient in fines. The use of low-activity admixtures is practised only to a very limited extent, and is of no particular advantage in a hot environment. Hence, this type of admixture is not further discussed.

3.1.2. Pozzolanic Admixtures

3.1.2.1. Pozzolanic Activity

Pozzolanic admixtures, or 'pozzolans', contain reactive silica (SiO_2), and sometimes also reactive alumina (Al_2O_3), which, in the presence of water, react with lime ($Ca(OH)_2$) and give a gel of calcium silicate hydrate (CSH gel) similar to that produced by the hydration of Portland cement. Accordingly, pozzolans are 'silicious or silicious and aluminous materials which, in themselves, possess little or no cementitious value but will, in a finely divided form and in the presence of moisture, chemically react with calcium hydroxide at ordinary temperatures to form compounds possessing cementitious properties' (ASTM C219). Such material are said to exhibit 'pozzolanic activity' and the chemical reactions involved are known as 'pozzolanic reactions'.

In the hydration of Portland cement (see section 2.3), a considerable amount of calcium hydroxide is produced. Hence, in mixtures made of a pozzolan and Portland cement, a pozzolanic reaction will take place due to the availability of lime. This availability of lime facilitates the replacement of some part of

Portland cement by pozzolans and explains why such an admixture can be used to produce pozzolan-based blended cements.

3.1.2.2. Classification

Generally, the pozzolans may be subdivided into natural and by-product materials. The former are naturally occurring materials, and their processing is usually limited to crushing, grinding and sieving. Such materials include volcanic ashes and lava deposits (e.g. volcanic glasses and volcanic tuffs) and are known, accordingly, as 'natural pozzolans'. Another type of natural pozzolan is diatomaceous earth, i.e. an earth which is mainly composed of the silicious skeletons of diatoms deposited from either fresh or sea water.

Naturally occurring clays and shales do not exhibit pozzolanic properties. However, when heat treated in the temperature range 600–900°C, such materials become pozzolanic and are referred to as 'burnt' or 'calcined pozzolans'. Strictly speaking, the latter are actually 'artificial pozzolans', but usually they are grouped together with natural pozzolans (ASTM C618).

As mentioned earlier, another group of pozzolans are by-product materials of some industrial process. The most common materials in this group are pulverised fly-ash (PFA) and condensed silica fume (CSF).

3.1.2.2.1. Pulverised fly-ash (PFA). Coal contains some impurities such as clays, quartz, etc. which, during the coal combustion, are fused and subsequently solidify to glassy spherical particles. Most of the particles are carried away by the flue gas stream and later are collected by electrostatic precipitators. Hence, as mentioned earlier, this part of the ash is known as fly-ash in the US, and pulverised fly-ash in the UK. The remaining part of the ash agglomerates to give what is known as 'bottom ash'.

Generally, fly-ash consists mostly of silicate glass containing mainly calcium, aluminium and alkalis. The exact composition, and the resulting properties of fly-ash, may vary considerably, and in this respect the CaO content is very important. Accordingly, fly-ashes are subdivided into two groups: low-calcium fly-ashes (CaO content less than 10%), and high-calcium fly-ashes (CaO content greater than 10%, and usually between 15 and 35%). This difference in CaO content is reflected in the properties of the fly-ashes. Whereas, for example, high-calcium fly-ashes are usually both pozzolanic and cementitious, low-calcium fly-ashes are only pozzolanic.

ASTM C618 classifies fly-ashes in accordance with their origin, namely, class F refers to fly-ashes which are produced from burning anthracite or

Table 3.1. Classification and Properties of Fly-Ash and Raw or Calcined Natural Pozzolan for Use as a Mineral Admixture in Portland Cement Concrete in Accordance with ASTM Standard C618-89a

Component/property	*Mineral admixture class*		
	N	*F*	*C*
Chemical requirements			
(1) Silica (SiO_2) + alumina (Al_2O_3) + iron oxide (Fe_2O_3), min (%)	70·0	70·0	50·0
(2) Sulphur trioxide (SO_3), max (%)	4·0	5·0	5·0
(3) Moisture content, max (%)	3·0	3·0	3·0
(4) Loss on ignition, max (%)	10·0	6·0[a]	6·0
(5) Available alkalies as Na_2O, max (%) (optional)	1·5	1·5	1·5
Physical requirements			
(1) Fineness, amount retained when wet-sieved on 45 μm (No. 325) sieve, max (%)	34	34	34
(2) Strength activity index with Portland cement:			
At 7 days, min (% of control)	75	75	75
At 28 days, min (% of control)	75	75	75
With lime, at 7 days, min (MPa)	5·5	5·5	5·5
(3) Water requirement, max (% of control)	115	105	105
(4) Soundness—autoclave expansion or contraction, max (%)	0·8	0·8	0·8
(5) Uniformity requirements: The specific gravity and fineness of individual samples shall not vary from the average established by the 10 preceding tests, or by all preceding tests if the number is less than 10, by more than:			
Specific gravity, max variation from average (%)	5	5	5
Percent retained on No. 325 (45 μm), max (variation percentage points from average)	5	5	5

[a]The use of class F pozzolan containig up to 12% loss of ignition may be allowed if either acceptable performance records or laboratory test results are made available.

bituminous coal, and class C refers to fly-ashes which are produced from burning lignite or sub-bituminous coal (Table 3.1). Usually the CaO content of class C fly-ashes is greater than 10%, and that of class F is lower. That is, the classification into low-calcium and high-calcium fly-ashes is essentially identical to that of ASTM C618 into F and C classes.

In addition to the CaO content, the properties of fly-ashes are determined, to a great extent, by their particle sizes and coal content. Generally, the finer the particles the greater the rate of the pozzolanic reaction, and the resulting development of strength. That is, coarser particles are not desirable explaining, in turn, the maximum imposed by the standards on the amount of fly-ash

Table 3.2. Classification and Properties of Fly-Ash for use as a Mineral Admixture in Portland Cement Concrete in Accordance with British Standard BS 3892, Part 1, 1982 and Part 2, 1984

Component/property	*BS 3892*		
	Part 1: 1982	*Part 2: 1984* *Grade*	
	–	*A*	*B*
(1) Loss on ignition, max (%)	7·0	7·0	12·0
(2) Magnesia (MgO), max (%)	4·0	4·0	4·0
(3) Sulphuric anhydride (SO_3), max (%)	2·5	2·5	2·5
(4) Moisture content, max (%)	0·5	0·5	0·5
(5) Fineness, amount retained on 45 μm sieve, max (%)	12·5	12·5–30	30–60
(6) Water requirement, percent of that required for Portland cement alone, max	95	–	–

retained on a No. 325 sieve (45 μm). In fact, particle size, as measured by the latter parameter, is used to classify fly-ashes in the British Standards (Table 3.2).

The coal content is measured by the loss of ignition. The presence of coal in the fly-ash is not desirable, mainly because it increases the water demand due to its great specific surface area. That is, the higher the coal content, the greater the amount of water which is required to impart a certain consistency to otherwise the same concrete mix. An increased amount of water adversely affects concrete properties and, thereby, explains the maximum imposed by the standards on water requirement, on the one hand, and the loss of ignition, on the other (Tables 3.1 and 3.2).

3.1.2.2.2. Condensed silica fume (CSF). CSF or, simply, microsilica, or silica fume, is an extremely fine by-product of the silicon metal and the ferrosilicon alloy industries, consisting mainly of amorphous silica (SiO_2) particles.

The silicon metal is produced by reducing quartz by coal at the temperature of about 2000°C. The reduction of the quartz is not complete and some SiO gas is produced. Part of this gas escapes into the air, is oxidised to SiO_2, and the latter is condensed to very small and spherical silica particles. Hence, the reference to CSF [3.3].

The most notable properties of microsilica are its very small particle size and high silica content. The average diameter of the microsilica particles is about 0·1 μm, resulting in a very high specific surface area of some 20 000 m^2/kg. That is, the size of the microsilica particles is two orders of magnitude smaller

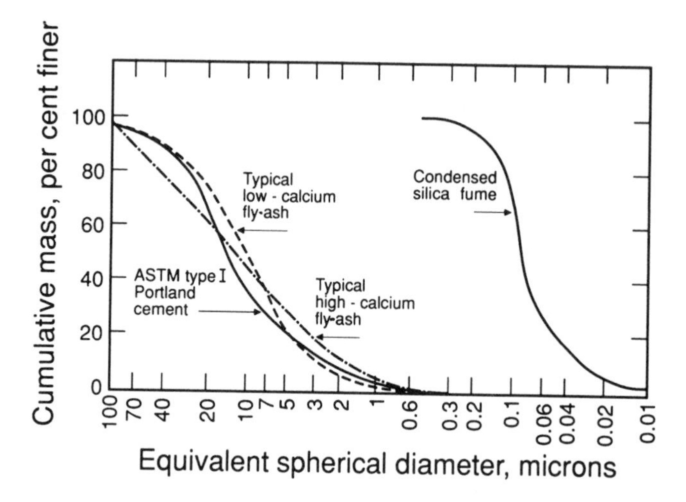

Fig. 3.1. Comparison of particle size distributions of Portland cement, fly-ash, and CSF. (Adapted from Ref. 3.2.)

than the size of the cement particles (average size 10 μm) or of fly-ash particles (Fig. 3.1). The silica content depends on the type of metal which is produced and varies, accordingly, from 84 to 98%.

The very high specific surface area, combined with the high silica content, accelerate the pozzolanic reactions, and thereby accelerate strength development (see section 3.1.2.3.4). In addition, the minute size of the silica fume particles produces a filler effect in the cement paste. This filler effect is schematically described in Fig. 3.2. On mixing with water, and for the same water to solids ratio, the initial porosity (i.e. the fractional volume occupied by the water) is the same in both systems considered. The very small silica fume particles, however, readily fill the spaces between the much coarser cement grains and, thereby, reduce the spacing between the solids. Hence, on subsequent hydration, the resulting capillary pores in the silica-fume-containing paste are much finer than the pores in the neat cement paste. That is, a more refined capillary pore system is brought about by incorporating silica fume in concrete mixes. Figure 3.3 presents experimental data which compare pore-size distributions in neat Portland cement and Portland cement plus silica fume pastes. It is clearly evident that the latter paste is characterised by a much finer pore system. This refinement in the pore system has important practical implications. It will be seen later that the lower permeability of silica-fume-containing concrete, and its associated improved durability, is attributable, partly at least, to the finer pore system which is brought about by the use of silica fume.

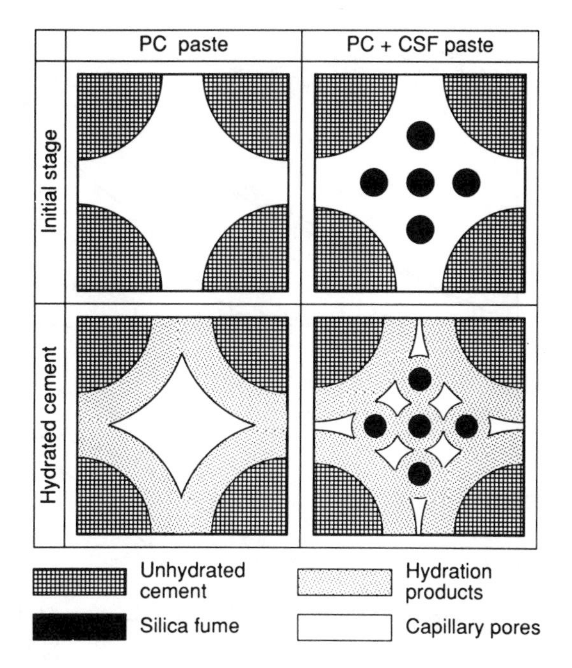

Fig. 3.2. Refinement of the pore-system in a cement paste due to the filler effect of silica fume.

The very high specific surface area of silica fume increases considerably the water demand of mortars and concretes, and this increase is greater the higher the silica fume addition (Fig. 3.4). In order to avoid such an increase, and its associated adverse effect on concrete properties, silica fume is always used with a water reducer, usually a high-range water reducer (see section 4.3.2). The specific water-reducing effect of such admixtures depends on many factors but it is usually more than enough to offset the increased water demand brought about by the use of silica fume.

3.1.2.3. Effect on Cement and Concrete Properties

The effect of pozzolans on the properties of Portland cement and concrete depends on the properties of the specific materials involved. Noting that even pozzolans of the same type may vary considerably, a general discussion of their effect is necessarily of a qualitative rather than of a quantitative nature. Accordingly, this is the nature of the following discussion whereas, in practice, the specific properties of the pozzolan in question must be considered.

3.1.2.3.1. Heat of hydration. Similarly to the hydration of Portland cements, the pozzolanic reactions result in the liberation of heat. The heat liberation due to

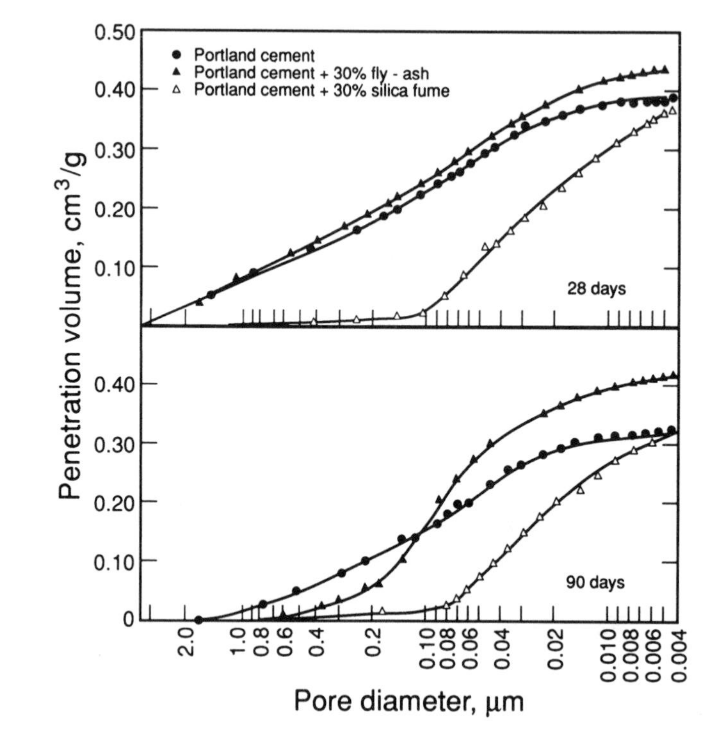

Fig. 3.3. Effect of replacing 30% of Portland cement (by absolute volume), with silica fume, or fly-ash, on pore-size distribution of the cement paste at the ages of 28 and 90 days. (Adapted from Ref. 3.4.)

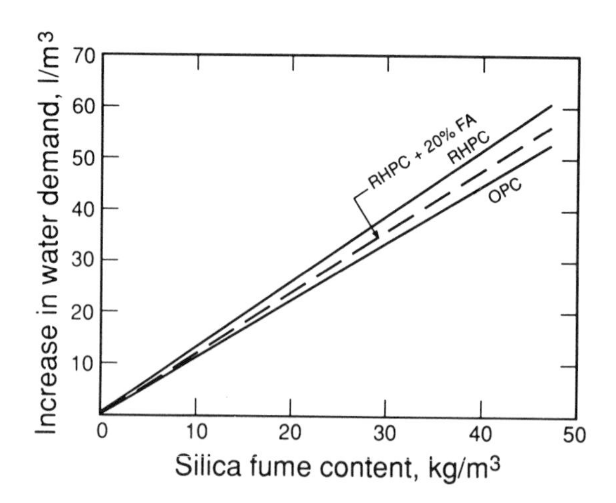

Fig. 3.4. Effect of silica fume content on water demand of concrete without a water-reducing agent. (Adapted from Ref. 3.5.)

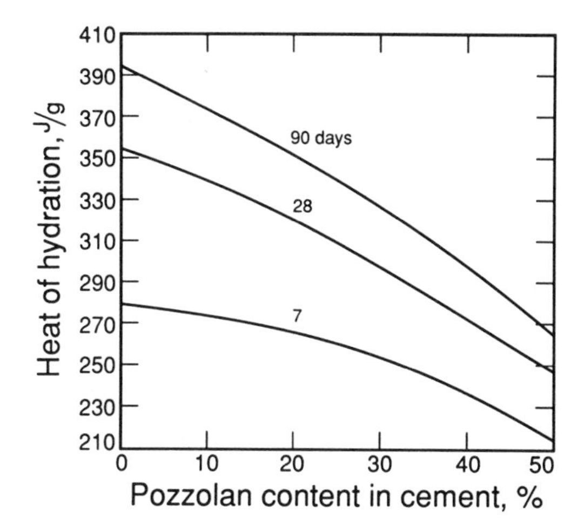

Fig. 3.5. Effect of partial replacement of Portland cement with an Italian natural pozzolan on the heat of hydration of the cement. (Adapted from Ref. 3.6.)

the latter reactions is less than that due to the hydration of Portland cement, and the rate of the pozzolanic reactions is lower than that of the hydration of Portland cement. Hence, replacing part of the Portland cement with a pozzolan would result in a cement with a lower heat of hydration, and the reduction in the heat of hydration would increase with the increase in the percentage of the Portland cement replaced by the pozzolan. The data presented in Fig. 3.5, which relate to an Italian natural pozzolan, clearly confirm these expected effects of pozzolanic admixtures on the heat of hydration of the cement. These effects are further confirmed by the data of Fig. 3.6, in which Portland cement was partly replaced by fly-ash (part A) and CSF (part B). Accordingly, it may be generally concluded that the partial replacement of Portland cement with a pozzolanic admixture results in a cement of a lower heat of hydration, and that such a cement may be used in lieu of low-heat Portland cement (see section 1.5.2).

The preceding conclusion with respect to the effect of CSF must be treated with some reservation. The very high specific surface area of the silica fume increases the rate of the pozzolanic reactions and thereby increases the rate of the resulting heat evolution. Hence, the heat of hydration of a cement containing silica fume may be higher than, say, its fly-ash-containing counterpart, and, perhaps, as high as, or even higher than, the heat of hydration of Portland cement. This expected effect is confirmed by the data of Fig. 3.7, but not by those of Fig. 3.6 where the silica fume was found to reduce the heat

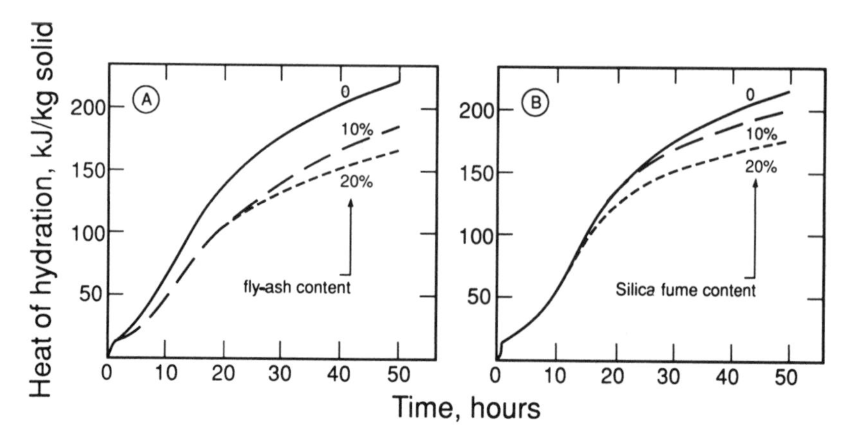

Fig. 3.6. Effect of partial replacement of Portland cement with (A) fly-ash, and (B) CSF, on the heat of hydration of the cement (cement pastes, water to solids ratio = 0·5). (Adapted from Ref. 3.7).

of hydration of the cement and, in this respect, the effects of both the silica fume and the fly-ash were essentially the same.

3.1.2.3.2. Microstructure. Replacing Portland cement with silica fume results in a finer pore system (Fig. 3.3). This effect of silica fume is attributable, partly at least, to the filler effect of the very small silica fume particles (Fig. 3.2). Such an effect, however, is not expected in other pozzolans which are characterised by a particle size similar to that of Portland cement.

The effect of replacing Portland cement with fly-ash on pore size distribution is also presented in Fig. 3.3. Accordingly, it can be seen that, at the age

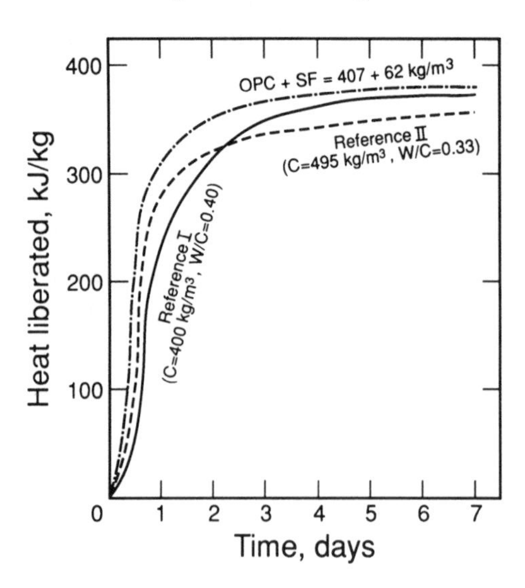

Fig. 3.7. Effect of partial replacement of Portland cement with condensed silica fume on the heat of hydration of the cement. (Adapted from Ref. 3.8.)

of 28 days, the fly-ash paste exhibited a somewhat greater porosity than its neat Portland cement counterpart, but the pore-size distribution of the two pastes was essentially the same. At the age of 90 days, although the porosity of the fly-ash paste remained higher than that of the neat Portland cement paste, its pore system became finer. Hence, it is usually accepted that the use of fly-ash is associated with a finer pore system, but not necessarily with a lower porosity. It may be realised that the finer pore system is reflected in lower permeability, provided the concrete is adequately cured. This aspect, however, is discussed later in the text (see section 9.2).

3.1.2.3.3. Calcium Hydroxide Content and pH of Pore Water. The consumption of calcium hydroxide due to the pozzolanic reactions is of practical importance when possible corrosion of the reinforcing steel of the concrete is considered. The presence of calcium hydroxide imparts to the pore water of the cement paste a high pH value of about 12·5, and such a high alkalinity protects the reinforcement against corrosion. This protection is lost, however, once the pH of the pore water drops below, say, 9, and it may be questioned if such a drop occurs due to the consumption of the calcium hydroxide by the pozzolanic reactions. That is, if this is really the case, the use of pozzolanic admixtures should be avoided, or even prohibited altogether, in reinforced concrete.

The effect of pozzolans and of other admixtures on possible corrosion of the reinforcing steel in concrete is discussed in some detail in Chapter 10. At this stage, however, it is enough to point out that the pozzolanic reactions lower only slightly the pH value of the pore water. This effect is demonstrated, for example, in Fig. 3.8 which relates to test data in which 15, 25 and 35% of Portland cement were replaced by two types of class F fly-ash. It can be seen that at the age of 150 days, and when fly-ash replaced 35% of the cement, the $Ca(OH)_2$ content was reduced by a factor greater than 2, whereas the pH value of the pore water dropped only slightly, i.e. from 12·97 to 12·72. Such a slight reduction was also reported by others when 30% of the cement was replaced by fly-ash [3.10].

A more significant reduction in the pH level was observed when silica fume was used to replace the cement, and particularly when the silica fume content was 30% (Fig. 3.9). However, when considering the more practical content of 10%, the reduction of the pH level remains insignificant.

3.1.2.3.4. Strength Development. The development of strength with time is brought about by the hydration of the cement because, as the hydration proceeds, the porosity of the cement paste decreases (see section 2.4). Similarly,

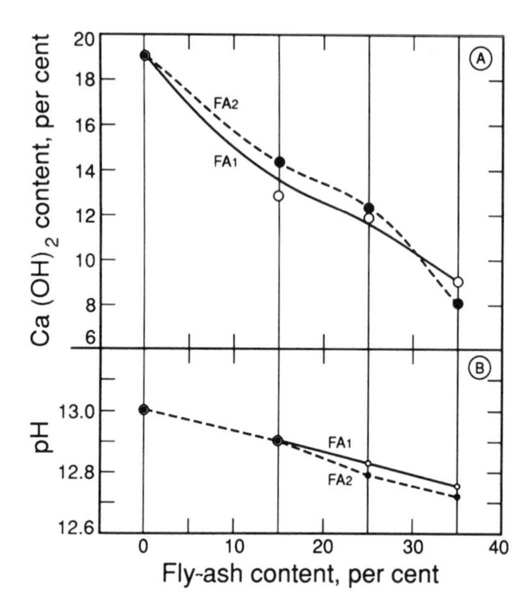

Fig. 3.8. Effect of fly-ash content on (A) $Ca(OH)_2$ content, and (B) pH value of the pore water, in Portland cement–fly-ash pastes at the age of 150 days. (Adapted from Ref. 3.9.)

the strength increases as the pozzolanic reactions proceed. The pozzolanic reactions are usually slower than the hydration of Portland cement and, consequently, the strength development of pozzolan–Portland cement blends is slower than the strength development of their unblended counterparts. Indeed, with the exception of blends in which silica fume is used, the early strength (i.e. for the first few weeks or even longer) of concretes made with a pozzolan-containing cement is lower than the strength of concretes made with unblended ordinary Portland cement (Fig. 3.10), and the higher the pozzolan content the greater the reduction in early strength [3.11, 3.12].

Although the preceding effects of pozzolans on early-age strength, have

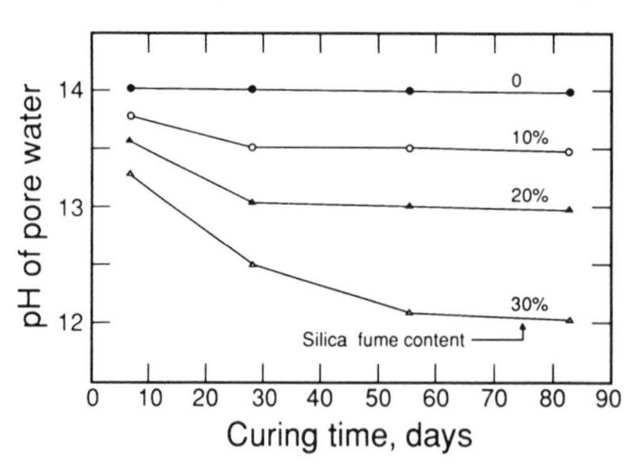

Fig. 3.9. Effect of silica fume content on the pH value of the pore water of cement pastes (W/(C + SF) = 0·50). (Adapted from Ref. 3.11.)

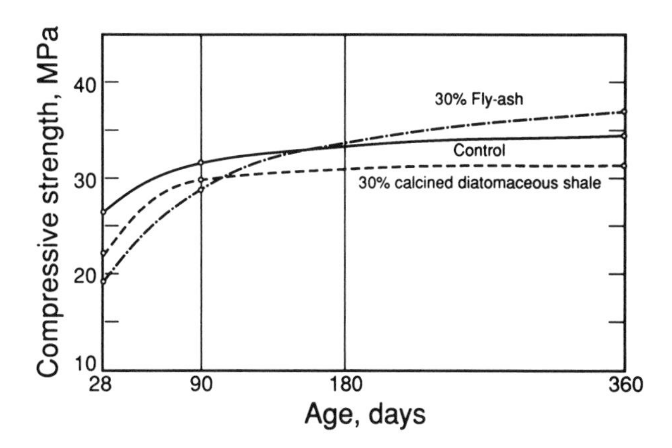

Fig. 3.10. Effect of replacing 30% of Portland cement by fly-ash, or by calcined diatomaceous earth, on concrete strength. (Adapted from Ref. 3.13.)

been widely observed and recognised, there exists some conflicting data with respect to their effect on later age strength. The data of Fig. 3.10, for example, indicate that replacing 30% of Portland cement by fly-ash produces a higher later age strength than that produced by the unblended cement, but not when replaced by the same amount of calcined diatomaceous shale. Yet, other data clearly indicate that the use of fly-ash is associated with both lower early and later age strengths (Fig. 3.11). These apparently contradictory data may be attributed to possible differences in curing conditions and the type of fly-ash involved. It seems that in practice, however, unless data are available to the contrary, it should be assumed that pozzolan–Portland cement blends produce

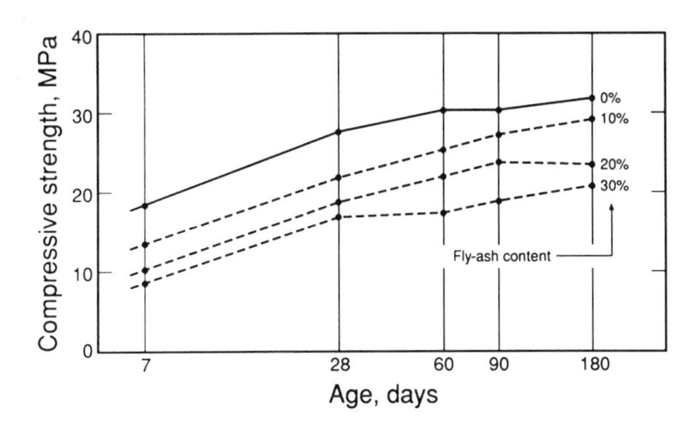

Fig. 3.11. Effect of replacing Portland cement by different amounts of fly-ash on concrete strength (OPC + FA = 320 kg/m^3, W/(C + FA) = 0·66, 7 days moist curing). (Taken from the data of Ref. 3.14.)

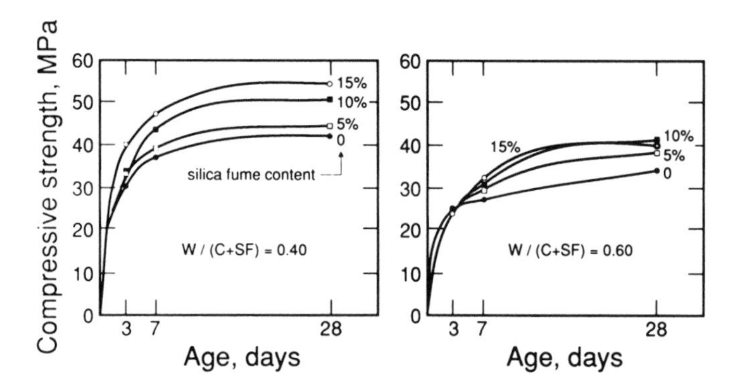

Fig. 3.12. Effect of replacing Portland cement by different amounts of silica fume on compressive strength of concrete. (Adapted from Ref. 3.15.)

lower strengths than their unblended counterparts, and particularly when the concrete is not cured for an extended period of time.

When silica fume is used to replace Portland, due to its high reactivity, concrete strength development is rather different from that observed when other pozzolans are used (Fig. 3.12). That is, the early-age strength is greater than that of unblended Portland cements, and the later-age strength is not only higher, but increases with the increase in the silica fume content as well. It may be noted that when other pozzolanic admixtures are used, a decrease in later age strength is observed when the admixture content is increased.

3.1.2.3.5. Other Properties. The preceding discussion deals with the effects of pozzolanic admixtures on some, but not on all, concrete properties. The admixtures effect on the remaining properties of concrete, such as volume changes and durability, requires some discussion of the properties in question before the effects of admixtures can be adequately treated. Hence, such treatment is presented in the relevant chapters.

3.1.3. Cementitious Admixtures

This type of admixture possesses hydraulic properties of its own and includes such materials as natural cements and hydraulic lime. However, by far the most common one is blast-furnace slag, or rather ground granulated blast-furnace slag. Hence, only this type of material is discussed hereafter.

3.1.3.1. Blast-Furnace Slag

Blast-furnace is a by-product of the pig iron industry, in which iron ores, mainly oxides of iron, are reduced to metallic iron.

The iron ores contain a certain amount of impurities, which are mainly SiO_2 and Al_2O_3. In order to separate these impurities from the melted iron, a certain amount of lime is added to the charge. The lime combines with the silica and the alumina, and the resulting molten slag, being much lighter than the molten iron, floats on top of the latter and is subsequently removed.

In order to give the slag hydraulic properties, it must be cooled rapidly, usually by quenching the liquid slag by water. In this process the molten slag, before being immersed in water, is broken up by water jets. Consequently, a slag having sand-size particles is produced, and is known, accordingly, as 'granulated' blast-furnace slag. Another, less common method, involves quenching by air with a limited amount of water. In this method, the slag is produced in the form of pellets and is known, accordingly, as 'pelletised' blast-furnace slag.

After being dried, the slag is ground together with Portland cement clinker to produce Portland blast-furnace slag cements (see section 3.2). Alternatively, it may be ground separately and the latter cements are produced by blending the pulverised slag with Portland cement. The use of slag on the building site as an admixture is, however, rather limited. Slag particles greater than 45 μm are barely reactive. Hence, the limitation imposed by ASTM C989, for example, on the amount of slag particles retained on the 45 μm sieve.

Blast-furnace slag is composed mainly of calcium, magnesium and alumina-silicates which are mostly, due to the rapid cooling of the slag, non-crystalline and glassy. In addition to particle size, both glass composition and content affect the reactivity of the slag. Usually a higher glass content, a higher combined content of lime (CaO), magnesia (MgO) and Al_2O_3 and a lower SiO_2 content, are associated with a higher reactivity. Moreover, the degree of the disorder of the glass structure affects significantly the reactivity of the slag and, generally speaking, a more disordered structure results in a more reactive slag.

The preceding effects of composition and glass content on the reactivity of the slag are reflected in BS 6699, 1986 by a minimum imposed on the contents of pure glass and glassy particles, the latter being particles which include both glass and crystalline material (Table 3.3). Similarly, the limitation of the composition is reflected in imposing a minimum on a so-called 'chemical

Table 3.3. Required Properties of Ground Granulated Blast-Furnace Slag for Use with Portland Cement in Accordance with BS 6699, 1986

Component/property	*Requirement*
(1) Fineness, min (m^2/kg)	275
(2) Glass content, percent of total particles, min	
Pure glass particles	40
Glassy particles	85
(3) Compressive strength of mortar made with 30:70 slag: Portland cement blend, min (MPa)	
At the age of 3 days	3
At the age of 28 days	22
(4) Setting time	
Initial (min)	$\geqslant 45$
Final (h)	<10
(5) Expansion, Le Chatelier test, 30:70 slag: Portland cement blend, min (mm)	10
Chemical composition	
(1) Insoluble residue, max (%)	1·5
(2) Magnesia, max (%)	4·0
(3) Sulphur, as sulphide, max (%)	2·0
(4) Loss of ignition, max (%)	3·0
(5) Manganese, expressed as Mn_2O_3, max (%)	2·0
(6) Chemical modulus ($CaO + MgO + Al_2O_3$) to SiO_2 ratio, min	1·0
(7) CaO to SiO_2 ratio, max	1·4
(8) Moisture content, max (%)	1·0

modulus' which is defined by the following ratio:

$$\text{Chemical modulus} = \frac{(CaO) + (MgO) + (Al_2O_3)}{(SiO_2)} \tag{3.1}$$

in which the bracketed formulae refer to the percentage of the particular oxide by mass in the slag. In accordance with BS 6699, 1986 the chemical modulus must be not less than one and the CaO to SiO_2 ratio not more than 1·4 (Table 3.3). No such limitations are imposed by the relevant ASTM Standard C989 (Table 3.4), which specifies the required reactivity of the slag more directly by a so-called 'slag activity index', which is defined by the following ratio:

$$\text{Slag activity index (\%)} = (SP/P) \times 100 \tag{3.2}$$

in which SP is the average compressive strength of a slag-cement reference mortar, i.e. a mortar prepared from a blend made of 50% slag and 50% Portland cement by mass; and P is the average compressive strength of otherwise the same mortar made of the unblended Portland cement.

Table 3.4. Classification and Required Properties of Ground Granulated Blast-Furnace Slag for use in Concrete and Mortars in Accordance with ASTM C989–89

Component/property	*Requirement*	
Chemical requirements		
(1) Sulphide sulphur (S), max (%)	2·5	
(2) Sulphate ion reported as SO_3, max (%)	4·0	
Physical requirements		
(1) Amount retained when wet-sieved on a 45 μm (No. 325) sieve, max (%)	20	
(2) Specific surface by air permeability, Method ASTM C204	No limits	
	Average of last five consecutive tests	*Any individual sample*
(3) Slag activity index, min (%)		
7 days index		
Grade 80	–	–
Grade 100	75	70
Grade 120	95	90
28 days index		
Grade 80	75	70
Grade 100	95	90
Grade 120	115	110

Granulated blast-furnace slag is actually a latent hydraulic binder and, in the absence of a suitable activator, its hydration is rather slow and of no practical use. The activation of slags may be brought about by strong alkalis, such as NaOH, KOH and $Ca(OH)_2$, and sulphates, such as gypsum ($CaSO_4$). Portland cement contains gypsum (see section 1.3.1), and some alkalis (see section 1.3.4). Moreover, a substantial amount of lime is produced as a result of the cement hydration. Hence, Portland cement can be used as an activator of blast-furnace slag and, indeed, is used in the production of Portland–blast-furnace slag cements (see section 3.2).

The hydration products of granulated blast-furnace slag are essentially similar to those produced by the hydration of Portland cement with the exception that no calcium hydroxide is produced.

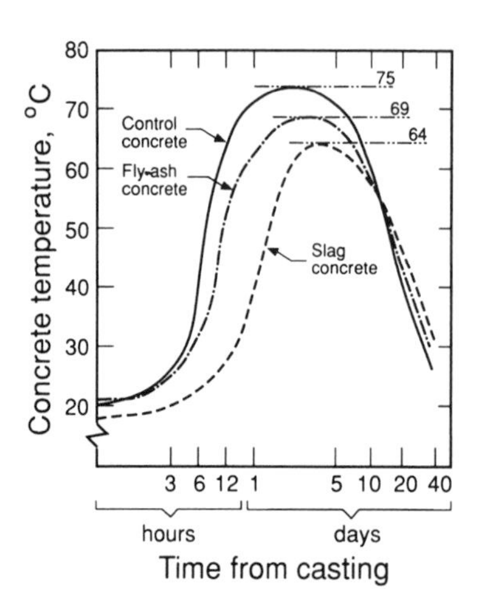

Fig. 3.13. Temperature rise measured in mass concrete. Control mix contained 400 kg/m³ ordinary Portland cement (OPC). In the fly-ash concrete 30% of the OPC was replaced by fly-ash and in the slag concrete 75% of the OPC was replaced by blast-furnace slag. (Adapted from Ref. 3.16.)

3.1.3.2. Effect on Cement and Concrete Properties

Similarly to the effect of the pozzolanic admixtures (section 3.1.2.3), the effect of blast-furnace slag on the properties of cement and concrete depends, to a great extent, on the specific properties of the slag involved. Hence, the following discussion is of a qualitative, rather than of a quantitative nature, and when the need arises, the properties of the slag involved must be determined and considered accordingly.

3.1.3.2.1. Heat of Hydration. The rate of slag hydration, and consequently the resulting heat evolution, is slower than the rate of hydration of Portland cement. Hence, replacing part of Portland cement with slag results in a cement with a lower heat of hydration and, consequently, the temperature rise in a concrete made of such a cement is lower than the temperature rise in otherwise the same concrete made of the same unblended Portland cement. This effect of substituting slag for Portland cement is demonstrated in Fig. 3.13, which presents the temperature rise which was recorded in mass concrete (i.e. foundations 4·5 m deep) made with and without replacing some of the Portland cement with slag or fly-ash. The effect of both the fly-ash and the slag, in controlling the temperature rise, is quite obvious and such a replacement may be considered, therefore, when a low-heat cement is required.

3.1.3.2.2. Microstructure. Experimental data, relating to the effect of slag content on total porosity of the cement paste, are presented in Fig. 3.14. It is clearly evident that porosity decreases with the increase in the slag content and, indeed, this effect on porosity has been suggested to explain the improved durability of concrete made with slag–Portland cement blends [3.17]. Such a decrease in porosity was not always observed and, in fact, an increase in the

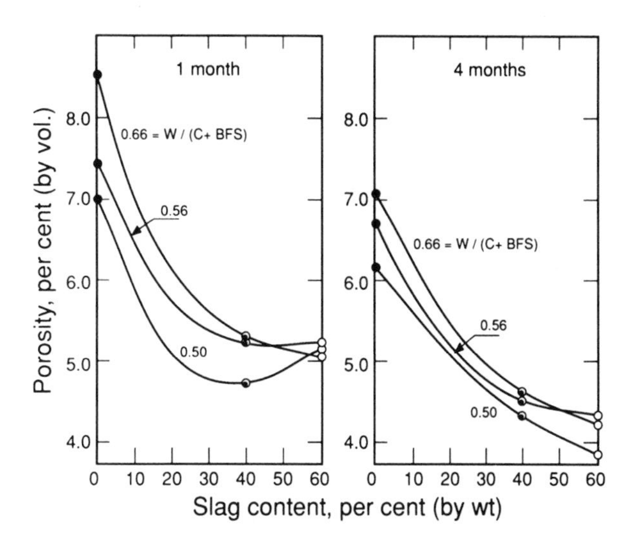

Fig. 3.14. Effect of slag content and water to cement (W/C) ratio on porosity of the cement paste. Pore-size from 30 to 7500 μm. (Adapted from Ref. 3.17.)

total pore volume, rather than a decrease, has been found with large additions of granulated blast-furnace slag (Fig. 3.15). Hence, the decreased permeability and the improved durability of concretes made of slag–Portland cement blends were attributed to the resulting finer pore system rather than to porosity as such [3.19, 3.20].

3.1.3.2.3. Strength Development. It was pointed out earlier that the rate of slag hydration is slower than that of Portland cement. Hence, a slower rate of heat evolution (Fig. 3.13) and a slower rate of strength development, are to be

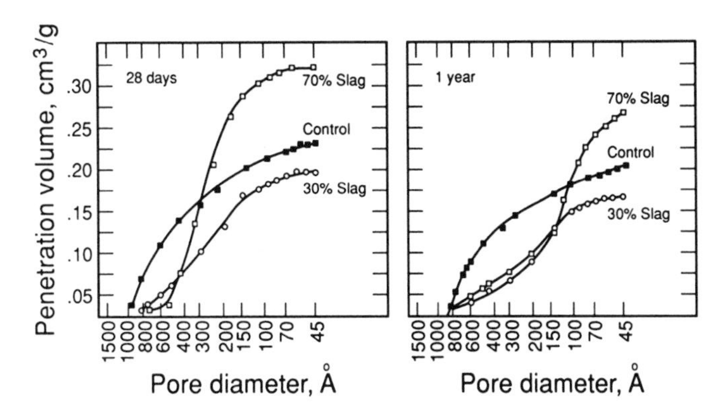

Fig. 3.15. Pore-size distribution of hydrated cements containing 30 or 70% granulated blast-furnace slag. (Adapted from Ref. 3.18.)

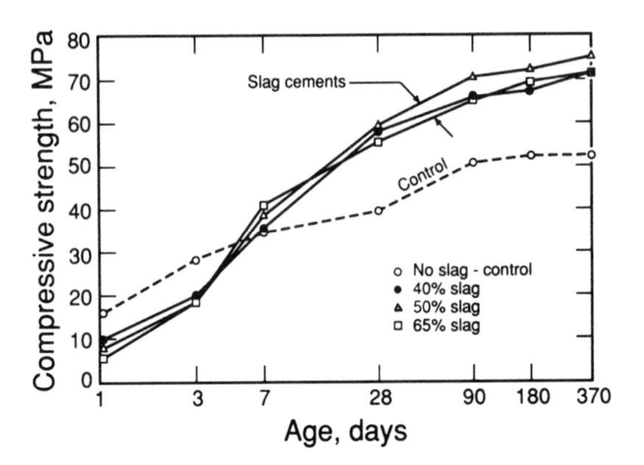

Fig. 3.16. Effect of slag content on compressive strength development in moist-cured mortars. (Taken from the data of Ref. 3.21.)

expected. This is usually the case and, indeed, the early strength of slag cement concrete is lower than the strength of its Portland cement counterpart. Later-age strength, however, in a well-cured concrete, may be comparable and even higher than that of otherwise the same concrete made from ordinary Portland cement (Fig. 3.16). Although this is generally the case, an exception may occur (Fig. 3.17) and, in practice, the possibility of such an occurrence must be considered.

3.1.3.2.4. Other Properties. The preceding discussion is limited to the effect of ground blast-furnace slag on certain concrete properties. Of course, the slag affects the remaining properties of concrete, such as volume changes and

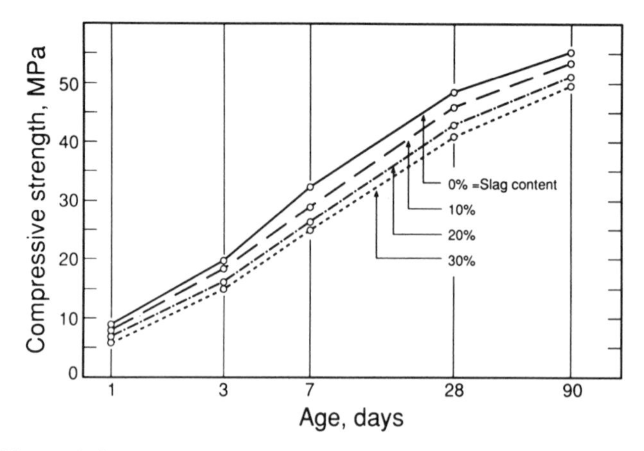

Fig. 3.17. Effect of slag content on compressive strength of concrete ($C + S = 330\,kg/m^3$, $W/(C + S) = 0{\cdot}61$). (Adapted from Ref. 3.22.)

durability, as well. As in the case of fly-ash, these effects are discussed later in the text, together with the relevant properties in question.

3.1.4. Summary

It must be realised that the classification of mineral admixtures into pozzolanic and cementitious materials is not always clear cut. Blast-furnace slag, for example, in addition to its cementitious properties, may possess also pozzolanic properties. On the other hand, high-calcium fly-ash may possess, in addition to its pozzolanic properties, cementitious properties as well. This aspect, together with a more detailed classification of mineral admixtures in accordance with the nature of their reactivity, can be found in Ref. 3.2.

3.2. BLENDED CEMENTS

3.2.1. Definition and Classification

Blended cement is 'an hydraulic cement consisting essentially of an intimate and a uniform blend of granulated blast-furnace slag and hydrated lime, or an intimate and uniform blend of Portland cement and granulated blast-furnace slag, Portland cement and pozzolan, or Portland blast-furnace slag cement and pozzolan, produced by intergrinding Portland cement clinker with the other materials, or by blending Portland cement with the other materials, or a combination of intergrinding and blending' (ACI Committee 116) [3.23].

In view of the preceding definition, it may be realised that quite a few blended cements can be produced by using different materials and blending proportions. Indeed, this is the case and ASTM Standard C595, for example, recognises the following types of blended cements.

(1) Portland–blast-furnace slag cement is a blended cement in which the slag content varies between 25 and 70% by weight of the cement. This type of cement, designated type IS, is intended for use in general concrete construction.

(2) Slag-modified Portland cement is a blended cement in which the slag content is less than 25% by weight of the cement. This type of cement, designated type I(SM), is intended for use for general concrete construction when the special characteristics attributed to the larger quantities of slag in the cement are not desired.

(3) Portland–pozzolan cement is a blended cement which contains a pozzolan. This definition covers two types of cement, namely, Portland–pozzolan cement, designated type IP, which contains 15–40% pozzolan and is intended for use in general concrete construction, and a similar cement, designated type P, which is intended for use in concrete construction where high strengths at early ages are not required. No limitations are imposed on the pozzolan content of type P cement.
(4) Pozzolan–modified Portland cement is a blended cement in which the pozzolan content is less than 15% by weight of the cement. This type of cement, designated type I(PM), is intended for use in general concrete construction.
(5) Slag cement is a blended cement in which the slag content is at least 70% by weight of the cement. This type of cement, designated type S, is used in combination with Portland cement in making concrete and with hydrated lime in making masonry mortar.

Similarly, the British standards recognise the following types of blended cements.

(1) Portland–blast-furnace cement is a cement in which the slag content does not exceed 65% of the total weight of the cement (BS 146, Part 2).
(2) Low-heat Portland–blast-furnace cement is a cement in which the slag content varies between 50 and 90% by weight of the cement (BS 4246, part 2).
(3) Portland–pulverised fuel-ash cement is a cement in which the ash content varies between 15 and 35% by weight of the cement (BS 6588, Part 2).
(4) Pozzolanic cement with pulverised fuel-ash as pozzolana is a cement in which the ash content varies between 35 and 50% of the total weight of the cement (BS 6610).

The preceding classifications are summarised in Tables 3.5 and 3.6 in accordance with British and ASTM standards, respectively.

3.2.2. Properties

Some difference may be expected in the properties of a cement produced by intergrinding Portland cement clinker with a mineral admixture, and those of

Table 3.5. Classification and Properties of Blended Cements in Accordance with British Standards

Component/properties	*Portland–blast-furnace slag cement*		*Portland–pulverised fuel ash cement*	
	BS 146, Part 2, 1973	*BS 4246, Part 2, 1974*	*BS 6588, 1985*	*BS 6610 1985*
Mineral composition				
(1) Admixture	Slag	Slag	Fly-ash	Fly-ash
(2) Content (%)	max 65	50–90	15–35	35–50
Chemical composition				
(1) Insoluble residue, max (%)	1·5	1·5	–	–
(2) Magnesia (MgO), max (%)	7·0	9·0	4·0	4·0
(3) Sulphuric anhydrite (SO_3), max (%)	3·0	3·0	3·0	3·0
(4) Loss on ignition, max (%)				
In temperate climates	3·0	–	4·0	4·5
In tropical climates	4·0	–	5·0	5·5
Physical properties				
(1) Compressive strength—concrete cubes min (MPa) at the age of				
3 days	8	3	8	8
7 days	14	7	–	–
28 days	22	14	22	16
(2) Fineness, min (m^2/kg)	225	275	225	225
(3) Setting time				
Initial, min (min)	45	60	45	45
Final, max (h)	10	15	10	10
(4) Soundness, max expansion (mm)	10	10	10	10
(5) Heat of hydration, max (kJ/kg), at the age of:				
7 days	–	250	–	–
28 days	–	290	–	–

a cement produced by mixing the very same ingredients, separately ground, on the building site or in a cement works. In this respect, it must be realised that the effect of admixtures on the cement properties is discussed here in a qualitative way only, and in such a discussion the effect of the production method is rather limited. Hence, the effect of partial replacement of Portland cement by mineral admixtures on the properties of the resulting blend, is valid also for blended cements whether or not such cements are produced by

Table 3.6. Classification and Properties of Blended Cements in Accordance with ASTM C595–89

Component/property	*Portland–blast-furnace slag cement (IS)*	*Slag-modified Portland cement (I(SM))*	*Slag cement (S)*	*Portland–pozzolan cement (IP)*	*Pozzolan-modified Portland cement (I(PM))*	*Portland–pozzolan cement (P)*
Mineral composition						
(1) Admixture	Slag	Slag	Slag	Pozzolan	Pozzolan	Pozzolan
(2) Content (%)	25–70	< 25	> 70	15–40	< 15	–
Chemical requirements						
(1) Magnesium oxide (MgO), max (%)	–	–	–	5·0	5·0	5·0
(2) Sulphate (SO_3), max (%)	3·0	3·0	4·0	4·0	4·0	4·0
(3) Sulphide sulphur (S), max (%)	2·0	2·0	2·0	–	–	–
(4) Insoluble residue, max (%)	1·0	1·0	1·0	–	–	–
(5) Loss on ignition, max (%)	3·0	3·0	4·0	5·0	5·0	5·0
(6) Water-soluble alkalis, max (%)	–	–	0·03	–	–	–
Physical requirements						
(1) Fineness			–			
(2) Autoclave expansion, max (%)	0·50	0·50	0·50	0·50	0·50	0·50
(3) Autoclave contraction, max (%)	0·20	0·20	0·20	0·20	0·20	0·20
(4) Time of setting, Vicat test						
Initial set (min)	> 45	> 45	> 45	> 45	> 45	> 45
Final set (h)	< 7	< 7	< 7	< 7	< 7	< 7
(5) Air content of mortar (Method C185), volume, max (%)	12	12	12	12	12	12
(6) Compressive strength, min (MPa) at the age of						
3 days	12·4	12·4	–	12·4	12·4	–
7 days	19·3	19·3	4·1	19·3	19·3	10·3
28 days	24·1	24·1	10·3	24·1	24·1	20·7

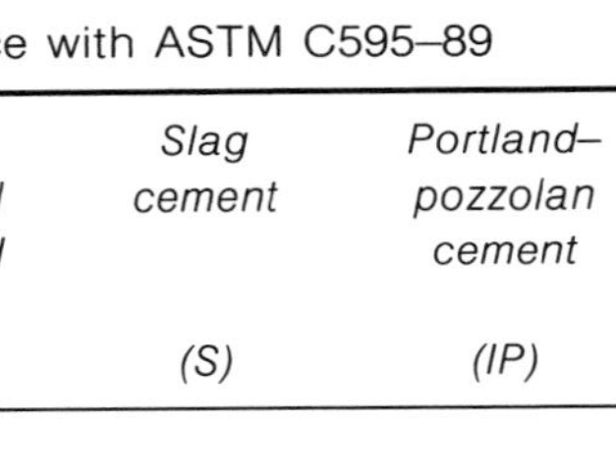

(7) Heat of hydration, max (kJ/kg) at the age of						
7 days	293	293	–	293	293	251
28 days	335	335	–	335	335	293
(8) Water requirement, max weight % of cement	–		–	–	–	64
(9) Drying shrinkage, max (%)	–	–	–	–	–	0·15
(10) Mortar expansion, max (%) at the age of						
14 days	0·020	0·020	0·020	0·020	0·020	0·020
8 weeks	0·060	0·060	0·060	0·060	0·060	0·060

intergrounding, or by intimately mixing their constituents. That is, the preceding sections (3.1.2.3 and 3.1.3.2) are applicable to blended cements as well. The required properties of the latter cements, in accordance with the British and ASTM standards, are summarised in Tables 3.5 and 3.6, respectively.

3.3. SUMMARY AND CONCLUDING REMARKS

Mineral admixtures are finely divided materials which are incorporated in the concrete mix in relatively large amounts (i.e. usually 15% or more, by weight of the cement), either as an addition or as partial replacement of the cement. Similarly, admixtures are used in the production of blended cements, i.e. cements which are an intimate blend of an admixture and Portland cement.

Mineral admixtures are subdivided into low activity (sometimes 'inert'), pozzolanic and cementitious materials. The pozzolans are further subdivided to natural, either raw or calcined, and by-product materials. Pulverised fly-ash (PFA), or simply fly-ash (FA) in the US, and condensed silica fume (CSF), or simply silica fume (SF), are by-product materials.

Cementitious admixtures possess hydraulic properties of their own. The most common material in this class is, by far, granulated blast-furnace slag which is a by-product of the iron industry. In order to impart the slag hydraulic properties, it must be cooled rapidly, and this is usually done by water quenching of the liquid slag.

Blast-furnace slag is actually a latent hydraulic binder and requires a suitable activator in order to be of practical use. Suitable activators include strong alkalis, such as NaOH, KOH and $Ca(OH)_2$ and sulphates, such as gypsum ($CaSO_4$). Portland cement is a suitable activator and as such is used in the production of blended cements known as Portland–blast-furnace slag cements.

Mineral admixtures, with the exception of CSF, are used to produce 'blended' cements, i.e. cements consisting essentially of an intimate mix of Portland cement and either blast-furnace slag, or a pozzolan (including fly-ash) or both. The properties of the blended cements depend on the specific properties of the admixture and the Portland cement used, and on their blending proportions.

REFERENCES

3.1. RILEM Committee 73–SBC, Siliceous by-products for use in concrete: Final report. *Mater. Struct.*, **21**(121) (1988), 69–80.

3.2. Mehta, P.K., Pozzolanic and cementitious by-products as mineral admixtures for concrete—A critical review. In *Fly Ash, Silica Fume, and Other Mineral By-Products in Concrete* (ACI Spec. Publ. SP-79, Vol. I), ed. V.M. Malhotra. ACI, Detroit, MI, USA, 1983, pp. 1–46.

3.3. ACI Committee 226, Silica fume in concrete. *ACI Mater. J.*, **84**(2) (1987) 158–66.

3.4. Mehta, P.K. & Gjorv, O.E., Properties of Portland cement concrete containing fly ash and condensed silica fume. *Cement Concrete Res.*, **12**(5) (1982), 587–95.

3.5. Sellevold, E.J. & Radjy, F.F., Condensed silica fume (microsilica) in concrete —Water demand and strength development. In *Fly Ash, Silica Fume, Slag and Other Mineral By-Products in Concrete* (ACI Spec. Publ. SP-79, Vol. II), ed. V.M. Malhotra. ACI, Detroit, MI, USA, 1983, pp. 677–94.

3.6. Massazza, F. & Costa, M., Aspects and pozzolanic activity and properties of pozzolanic cements. *Il Cemento*, **76**(1) (1979),507–18.

3.7. Meland, I., Influences on condensed silica fume and fly ash on the heat evolution in cement pastes. In *Fly Ash, Silica Fume, Slag and Other Mineral By-Products in Concrete* (ACI Spec. Publ. SP-79, Vol. II), ed. V.M. Malhotra. ACI, Detroit, MI, USA, 1983, pp. 665–76.

3.8. Bentur, A. & Goldman, A., Curing effects, strength and properties of high-strength silica fume concretes. *J. Mater. Civ. Engng*, **1**(1) (1988), 46–58.

3.9. Leonard, S. & Bentur, A., Improvement of the durability of glass fiber reinforced cement using blended cement matrix. *Cement Concrete Res.*, **14**(5) (1984), 717–28.

3.10. Diamond, S., Effects of two Danish fly ashes on alkali contents of pore solutions of cement–fly ash pastes. *Cement Concrete Res.*, **11**(3) (1981), 383–94.

3.11. Page, C.L. & Vennesland, O., Pore solution composition and chloride binding capacity of silica–fume cement pastes. *Mater. Struct.*, **16**(91) (1983), 19–25.

3.12. Mehta, P.K., Studies on blended Portland cements containing santorin earth. *Cement Concrete Res.*, **11**(4) (1981), 507–18.

3.13. Higginson, E.G., Mineral admixtures. In *Significance of Tests and Properties of Concrete and Concrete Making Materials* (ASTM Spec. Tech. Publ. No. 169A). ASTM, Philadelphia, PA, USA, 1966, pp. 543–55.

3.14. Jaegermann, C. & Sikuler, Y., Effect of curing regime and temperature on strength development of fly ash concrete. Research Report 017-396, Building Research Station, Technion—Israel Institute of Technology, Haifa, 1987 (in Hebrew with an English synopsis).

3.15. Malhotra, V.M., Mechanical properties and freezing and thawing resistance of non-air entrained and air entrained condensed silica fume concrete using ASTM test C666, procedures A and B. *Proc. Sec. Intern. Conf. on Fly Ash, Silica Fume, Slag and Natural Pozzolans in Concrete* (ACI Spec. Publ. SP-91), Madrid, Spain 1986. ed. V.M. Malhotra. Detroit, pp. 1069–94.

3.16. Bamforth, P.B., In situ measurement of the effect of partial Portland cement replacement using either fly ash or ground granulated blastfurnace slag on the performance of mass concrete. *Proc. Inst. Civ. Engng*, **69** (1980), 777–800.

3.17. STUVO, *Concrete on Hot Countries*. The Dutch member group of FIP, Delft, The Netherlands.

3.18. Mehta, P.K., Sulfate resistance of blended Portland cements containing pozzolans and granulated blast furnace slag. In *Proc. 5th Intern. Symp. on Concrete Technology*. Monterrey, Mexico, 1981, ed. V.M. Malhatra. CANMET, Ottawa, pp. 35–50.

3.19. Manmohan, D. & Mehta, P.K., Influence of pozzolanic, slag and chemical admixtures on pore-size distribution and permeability of hardened cement pastes. *Cement Concrete and Aggregates*, **3**(1) (1981), 63–7.

3.20. Feldman, R.F., Pore structure formation during hydration of fly ash and slag cement blends. In *Cement and Concrete, Proc. Symp. N, Materials Research Society*, ed. S. Diamond. Materials Research Society, Philadelphia, 1981, pp. 124–33.

3.21. Hogan, F.J. & Meusel, J.W., Evaluation for durability and strength development of a ground granulated blast furnace slag. *Cement Concrete and Aggregates*, **3**(1) (1981), 40–51.

3.22. Ravina, D., Properties of cement and concrete containing blastfurnace slag. Research Report 017-433, National Building Research Institute, Technion—Israel Institute of Technology, Haifa, 1990 (in Hebrew).

3.23. ACI Committee 116, *Cement and Concrete Terminology* (ACI 116R-85). In ACI Manual of Concrete Practice, Part 1. ACI, Detroit, MI, USA, 1990.

Chapter 4
Workability

4.1. INTRODUCTION

The 'workability' of concrete may be defined as 'the property determining the effort required to manipulate a freshly mixed quantity of concrete with a minimum loss homogeneity' (ASTM C125). In this definition the term 'manipulate' is meant to include all the operations involved in handling the fresh concrete, namely, transporting, placing, compacting and also, in some cases, finishing. In other words, workability is that property which makes the fresh concrete easy to handle and compact without an appreciable risk of segregation.

The workability may be defined somewhat differently and, indeed, other definitions have been suggested. Nevertheless, and regardless of the exact definition adopted, it may be realised that the workability is a composite property and, as such, cannot be determined quantitatively by a single parameter. In practice, however, such a determination is required and, strictly speaking, common test methods (slump, Vebe apparatus) actually determine the 'consistency' or the 'compactability' of the fresh concrete rather than its 'workability'. In practice, however, workability and consistency are usually not differentiated.

Generally, the workability is essentially determined by the consistency and cohesiveness of the fresh concrete. That is, in order to give the fresh concrete the desired workability, both its consistency and cohesiveness must be controlled. The sought-after cohesiveness is attained by proper selection of mix proportions using one of the available mix-design procedures [4.1, 4.2]. In other words, once cohesiveness is attained, the workability is further controlled by the consistency alone. This is usually the case and in practice, indeed, workability is controlled by controlling the consistency of the mix. Hence, the sometimes indiscriminate reference to 'consistency' and 'workability', as well as the use of consistency tests such as the slump, or the Vebe tests to control workability (BS 1881, Parts 102, 103 and 104). In this respect it is further assumed that a stiffer mix is less workable than a more fluid one, and vice versa. This assumption, however, is not always true, because a very wet mix may exhibit a marked tendency to segregate, and as such is, therefore, of a poor workability.

4.2. FACTORS AFFECTING WATER DEMAND

4.2.1. Aggregate Properties

The consistency of the fresh concrete is controlled by the amount of water which is added to the mix. The amount of water required (i.e. the 'water demand' or 'water requirement') to produce a given consistency depends on many factors such as aggregate size and grading, its surface texture and angularity, as well as on the cement content and its fineness, and on the possible presence of admixtures. The water wets the surface of the solids, separates the particles, and thereby acts as a lubricant. Hence, the greater the surface area of the particles, the greater the amount of water which is required for the desired consistency, and vice versa. Similarly, when a greater amount of mixing water is used, the separation between the solid particles is increased, friction is thereby reduced, and the mix becomes more fluid. The opposite occurs when a smaller amount of water is added, i.e. friction is increased bringing about a stiffer mix. Hence, the sometimes synonymous use of 'wet' and 'fluid' mixes on the one hand, and the use of 'dry' and 'stiff' mixes, on the other.

It must be realised, however, that quantitatively the relation between the consistency and the amount of mixing water is not linear, but rather of an exponential nature. It can be generally expressed mathematically by the following expression:

$$y = CW^n$$

where y is the consistency value (e.g. slump etc.); W is the water content of

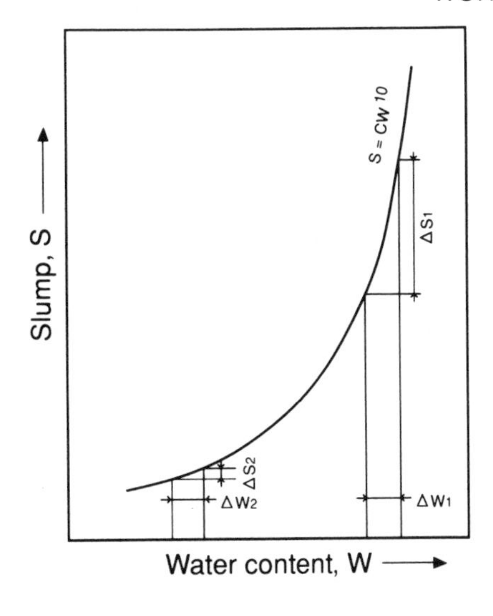

Fig. 4.1. Schematic representation of the relation between slump and the amount of mixing water. (Adapted from Ref. 4.3.)

the fresh concrete; C is a constant which depends on the composition of the mix, on the one hand, and the method of determining the consistency, on the other; n is also a constant which depends, again, on the method of determining the consistency but not on concrete composition. A graphical representation of this equation is given in Fig. 4.1 for $n = 10$.

It is clearly evident from Fig. 4.1 that the slump of the wetter mixes is more sensitive to changes in the amount of mixing water than the slump of the stiffer ones. In other words, a given change in the amount of mixing water ($\Delta W_1 = \Delta W_2$) causes a greater change in the slump of the wetter mixes than in the slump of the stiffer ones ($\Delta S_1 > \Delta S_2$).

Generally, the aggregate comprises some 70% by volume of the concrete, whereas the cement comprises only some 10%. Moreover, usually, the specific surfaces of the cements used in daily practice are more or less the same. Hence, in practice, excluding the effect of admixtures, the amount of water required to give the fresh concrete the desired consistency (usually specified by the slump), is estimated with respect to the aggregate properties only, i.e. with respect to aggregate size and shape. Size is usually measured by the parameter known as 'maximum size of aggregate', which is the size of the sieve greater than the sieve on which 15% or more of the aggregate particles are retained for the first time on sieving. In considering shape and texture, a distinction is made between 'crushed' and 'uncrushed' (gravel) aggregate. The particles of crushed aggregate are angular and of a rough texture whereas those of gravel aggregate, are round and smooth. Hence, the latter are characterised by a smaller surface area, and require less water than the crushed aggregate to produce a mix of a given consistency.

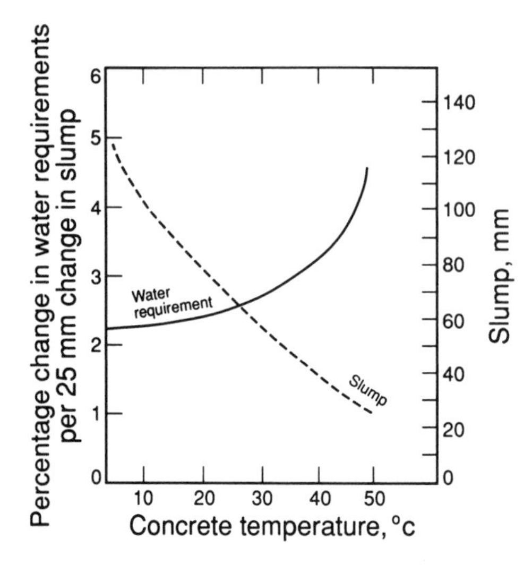

Fig. 4.2. Effect of concrete temperature on slump and amount of water required to change slump. Cement content of about 300 kg/m^3, types I and II cements, maximum size of aggregate 38 mm, air content of 4·5 ± 0·5%. (Adapted from Ref. 4.4.)

4.2.2. Temperature

It is well known that under hot weather conditions more water is required for a given mix to have the same slump, i.e. the same consistency. This is demonstrated, for example, in Figs 4.2 and 4.3, and it can be seen (Fig. 4.2) that, under the conditions considered, approximately a 25 mm decrease in slump was brought about by a 10°C increase in concrete temperature. Alternatively, it is indicated in Fig. 4.3 that the water demand increases by 6·5 kg/m^3 for a rise of 10°C in concrete temperature. An increase of 4·6 kg/m^3 for the same change in temperature has been reported by others [4.6].

The effect of temperature on water demand is mainly brought about by its

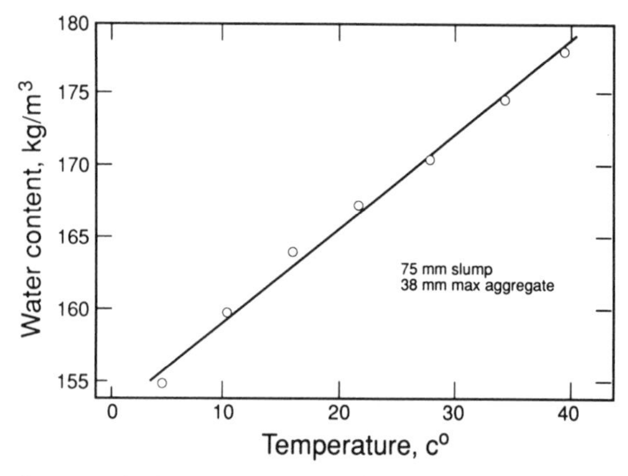

Fig. 4.3. Effect of concrete temperature on the amount of water required to produce 75 mm slump in a typical concrete. (Adapted from Ref. 4.5.)

effect on the rate of the cement hydration [4.7], and possibly also on the rate of water evaporation. The slump data of Figs 4.2 and 4.3 refer to the initial slump, i.e. to the slump determined as soon as possible after the mixing operation is completed. Nevertheless, some time elapses between the moment the water is added to the mix and the moment the slump is determined. The cement hydrates during this period and some water evaporates. Consequently, the mix somewhat stiffens and its slump, therefore, decreases. As the rates of hydration and evaporation both increase with temperature (see section 2.5.1), the associated stiffening is accelerated, and the resulting slump loss is, accordingly, increased. Hence, if a certain initial slump is required, a wetter mix must be prepared in order to allow for the greater slump loss which takes place when the concrete is prepared under higher temperatures. In other words, under such conditions, a greater amount of water must be added to the mix explaining, in turn, the increase in water demand with temperature. This important aspect of slump loss is further discussed in section 4.3 with particular reference to the role of temperature.

4.3. FACTORS AFFECTING SLUMP LOSS

4.3.1. Temperature

The fresh concrete mix stiffens with time and this stiffening is reflected in a reduced slump. Accordingly, this phenomenon is referred to as 'slump loss'. As already mentioned, this reduction in slump is brought about mainly by the hydration of the cement. Evaporation of some of the mixing water, and possible water absorption by the aggregates, may constitute additional reasons which contribute to slump loss. The formation of the hydration products removes some free water from the fresh mix partly due to the hydration reactions (i.e. some 23% of the hydrated cement by weight), and partly due to physical adsorption on the surface of the resulting hydration products (i.e. some 15% of the hydrated cement by weight). Again, more water may be removed by evaporation, and the resulting decrease in the amount of the free water reduces its lubricant effect. The friction between the cement and aggregates particles is increased, and the mix becomes less fluid, i.e. a slump loss takes place.

Once slump loss is attributed to the cement hydration and the evaporation of some of the mixing water, it is to be expected that a higher concrete

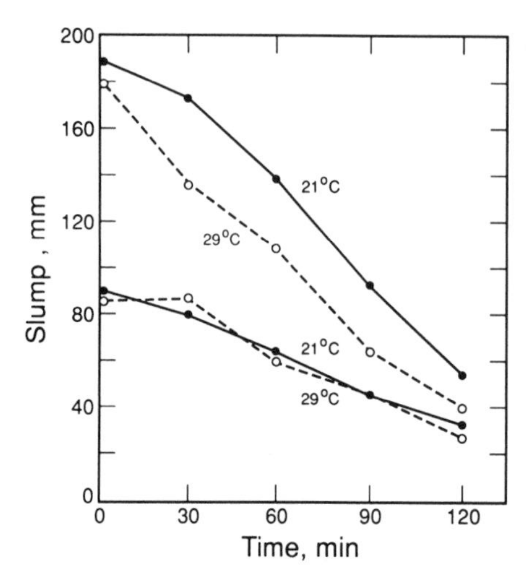

Fig. 4.4. Effect of temperature and initial slump on slump loss of concrete. (Taken from the data of Ref. 4.8.)

temperature will similarly accelerate the rate of slump loss. However, this expected effect of temperature is not always supported by experimental data. It can be seen from Fig. 4.4, for example, that the rate of slump loss was temperature dependent, at best only, in the wetter mixes (initial slump 180–190 mm) whereas in the stiffer mixes (initial slump of 90 mm) the rate remained the same and independent of temperature. Essentially, the same behaviour is indicated by the data of Fig. 4.5, i.e. the rate of slump loss in the wetter mixes (initial slump 205 mm) was greater at 32°C than at 22°C, whereas the rate in the stiffer mixes (initial slump 115–140 mm) remained virtually the

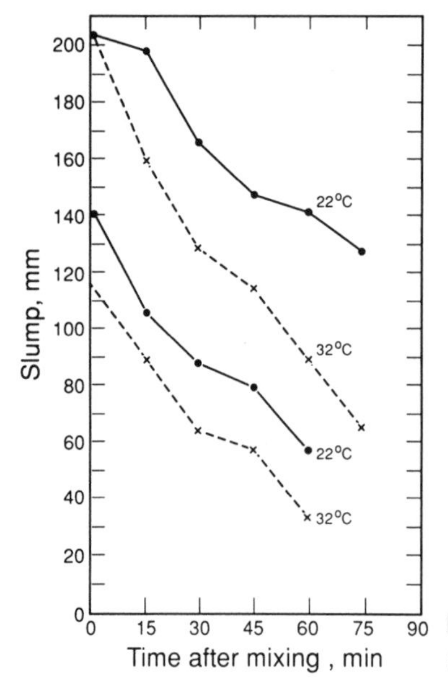

Fig. 4.5. Effect of temperature on slump loss. (Taken from the data of Ref. 4.9.)

same, i.e. the slump loss curves remained more or less parallel. This difference in the slump loss of wet and stiff mixes is attributable, partly at least, to the fact that the consistency of stiffer mixes is less sensitive to changes in the amount of mixing water than that of the wetter mixes (Fig. 4.1).

In view of the preceding discussion, it may be concluded that, in practice, the possible adverse effect of higher temperatures on consistency can be avoided, or at least greatly reduced, by the use of mixes characterised by a moderate slump, i.e. by a slump of, say, 100 mm. In principle, however, the slump loss of both wet and dry mixes must be temperature dependent because it is brought about by the hydration of the cement and the evaporation of some of the mixing water which, in turn, are both temperature dependent. Hence, it is generally accepted and, indeed, supported by the site experience, that slump loss of concrete is accelerated with temperature, and that this effect takes place not necessarily only in the wetter mixes. In fact, this accelerating effect of temperature on the rate of slump loss constitutes one of the main problems of concreting under hot weather conditions.

4.3.2. Chemical Admixtures

4.3.2.1. Classification

There are different types of chemical admixtures. ASTM C494, for example, recognises five types: water-reducing admixtures (type A), retarding admixtures (type B), accelerating admixtures (type C), water-reducing and retarding admixtures (type D), and water-reducing and accelerating admixtures (type E). These types of admixtures are sometimes collectively referred to as 'conventional admixtures'. Other types include air-entraining admixtures (ASTM C260) and high-range water-reducing admixtures (ASTM C1017), commonly known as superplasticisers. ASTM C1017 covers two types of superplasticiser which are referred to as plasticising (type 1), and plasticising and retarding admixtures (type 2). It must be realised that chemical admixtures are commercial products and, as such, although complying with the same relevant standards, may differ considerably in their composition and their specific effects on concrete properties.

4.3.2.2. Water-Reducing Admixtures

A water-reducing admixture is, by definition, 'an admixture that reduces the quantity of mixing water required to produce concrete of a given consistency' (ASTM C494). Generally, and depending on the cement content, type of aggregate, etc., and, of course, on the specific admixture involved, the actual water reduction varies between 5 and 15%. A greater reduction in water content cannot be achieved by using double or triple dosages because such an increased dosage may result in excessive air entrainment, an increased tendency to segregation and sometimes also in uncontrolled setting. The high-range water-reducing admixtures (superplasticisers) are a comparatively new breed of water-reducing admixtures which allow up to 25% reduction in the amount of mixing water without significantly affecting adversely the properties of the fresh and the hardened concrete (see section 4.3.2.4).

The accelerating effect of temperature on slump loss may be overcome by using, under hot weather conditions, a wetter mix than normally required under moderate temperatures. Increasing the amount of mixing water is the most obvious way to get such a mix. However, such an increase in mixing water is not desirable and, in any case, is applicable only up to a certain amount which, when exceeded, results in a mix with a high tendency to segregation. Consequently, increasing the amounts of mixing water may be a practical solution only under moderate conditions while under more severe conditions other means must be considered, such as the use of water-reducing admixtures. It must be realised, however, that the use of such admixtures may be associated, sometimes, with an increased rate of slump loss.

4.3.2.3. Retarding Admixtures

A retarding admixture is 'an admixture that retards the setting of the concrete' (ASTM C494). Accordingly, a water-reducing and retarding admixture combines the effects of both water-reducing and retarding admixtures, and as such delays setting and allows a reduction in the amount of mixing water as well. As has already been mentioned, admixtures types D and 2, in accordance with ASTM C494 and C1017, respectively, are such admixtures. Generally, these two types of admixtures are usually preferred for hot-weather concreting.

A retarding admixture slows down the hydration of the cement and thereby delays its setting. Hence, due to the slower rate of hydration, a smaller amount of water is combined with the cement at a given time. It is to be expected, therefore, that the corresponding slump loss in such a mix at the time

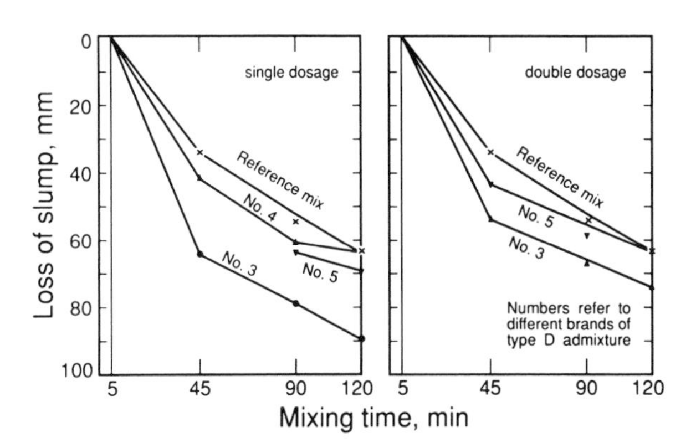

Fig. 4.6. Effect of water reducing and retarding admixtures on loss of slump. Type D admixtures, initial slump 95 to 115 mm, temperature 30°C. (Taken from the data of Ref. 4.10.)

considered will be smaller than in a mix made without an admixture. In other words, it is to be expected that the use of a retarding admixture would reduce the rate of slump loss and, therefore, may be useful in overcoming the accelerating effect of temperature. This expected effect, however, has not been confirmed by laboratory tests at least for conditions when transported concrete (ready-mixed) was considered, i.e. when the concrete was agitated from the time of mixing to the time of delivery.

The effect of type D admixtures on the slump loss of concrete subjected to 30°C is demonstrated in Fig. 4.6. It is evident that the presence of the admixtures, depending on their specific type and dosage, actually increased, rather than decreased, the rate of slump loss. This observation has been confirmed by many others [4.8, 4.11–4.14] and gives rise to the question whether or not this type of admixture may be recommended for use in hot weather conditions.

The increased rate of slump loss that was observed when some water-reducing admixtures were used, implies that the admixtures in question actually accelerated the rate of hydration. This, indeed, may be the case when type A admixtures are involved and, in fact, ASTM C494 allows the time of setting of concrete containing this type of admixture to be up to 1 h earlier than the time of setting of the control mix. That is, in this case, the admixture acts as an accelerator as well, and thereby causes a more rapid stiffening and a higher rate of slump loss. However, the increased slump loss observed when type D admixtures were used warrants some explanation because these types of admixtures do retard setting when tested in accordance with ASTM C494. The seemingly contradictory behaviour may be attributed to the difference in test conditions involved, i.e. while the increased slump loss was observed in concrete which was subjected, one way or another, to agitation, either continuously or periodically, the time of setting is determined on a concrete which remains undisturbed (ASTM C403).

Several theories have been advanced to explain the mechanism of retardation [4.15]. The adsorption theory suggests that the admixture adsorbs on the surfaces of the unhydrated cement grains, and thereby prevents the water from reacting with the cement. Another theory, the precipitation theory, suggests that the retardation is caused by the formation of an insoluble layer of calcium salts of the retarder on the hydration products. Agitating the concrete results in a grinding effect which, among other things, can be visualised as removing the adsorbed layer of the retarder or, alternatively, the precipitated layer of the calcium salts, whatever the case may be, from the surface of the cement grains. Hence, when the concrete is agitated, and particularly if the agitation takes place continuously and for long periods, the retarding mechanism fails to operate, and it is to be expected that under such conditions a type D admixture will behave, in principle, similarly to type A. In fact, such similar behaviour was observed in laboratory tests [4.8, 4.10]. It follows that, in practice, when long hauling periods are involved, there is no real advantage in using a type D admixture, and to this end the use of type A will produce essentially the same effects. This may not be the case in non-agitated concrete where the retarding effect of the type D admixture is desirable because it delays setting and helps to prevent cold joints, etc.

It will be seen later (section 4.4.1) that, although the use of water-reducing (type A) or water-reducing and retarding admixtures (type D) are, in many cases, associated with a higher rate of slump loss, the use of such admixtures is beneficial, provided they are used primarily to increase the initial slump of the mix and not necessarily to reduce the amount of mixing water. When short delivery periods are involved, increasing the initial slump of the concrete may provide the answer to the increased slump loss due to temperature. This may not be the case for long hauling periods where retempering may be required. It will be seen later that, under such conditions, the use of the admixtures in question may prove to be beneficial (section 4.4.3).

4.3.2.4. Superplasticisers

It was mentioned earlier that the use of superplasticisers affects the consistency of the concrete mix to a much greater extent than the use of conventional water reducers, facilitating a reduction of up to, say, 25% in the amount of

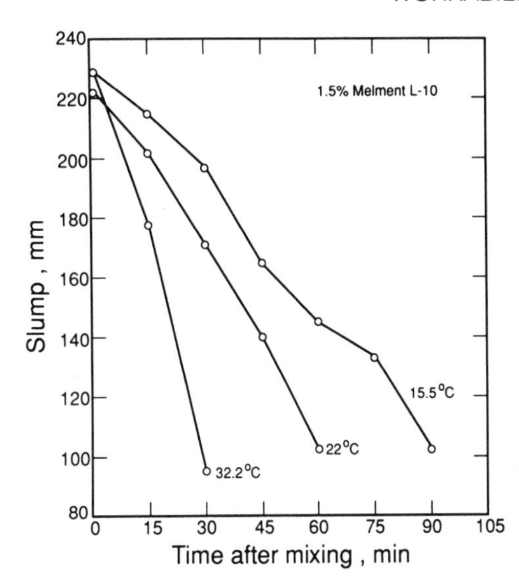

Fig. 4.7. Effect of temperature on slump loss of concrete made with a superplasticiser (1·5% Melment L-10). (Taken from the data of Ref. 4.16.)

mixing water without adversely affecting concrete properties. Consequently, when used to increase the fluidity of the mix, superplasticisers may increase slump from 50–70 mm to 200 mm or more, with the resulting mix remaining cohesive and exhibiting no excessive bleeding or segregation. Moreover, as the water to cement (W/C) ratio is not changed, the strength of the concrete remains virtually the same. Indeed, in such a way, superplasticisers are used to produce a so-called 'flowing concrete' which can be placed with little or no compaction at all, and is useful, for example, for placing concrete in thin and heavily reinforced sections. Flowing concrete may be useful also in hot weather conditions in order to overcome the adverse effect of the high temperatures on slump loss.

It must be realised, however, that the effect of superplasticisers on concrete consistency is comparatively short lived and, generally speaking, lasts only some 30–60 min from its addition to the mix, even under moderate temperatures. This period of time is much shorter under higher temperatures because the rate of slump loss of superplasticised mixes increases with temperature to an appreciable extent (Fig. 4.7). Moreover, similarly to concrete containing conventional water reducers (Fig. 4.6), the rate of slump loss in superplasticised concrete is usually, but not always, greater than the rate of slump loss in otherwise the same non-superplasticised concrete (Fig. 4.8). Apparently, new types of superplasticisers are now available which affect concrete consistency for longer periods, and thereby are more effective under hot weather conditions [4.18, 4.19]. In fact, superplasticiser C in Fig. 4.8 is such an admixture. It can be seen that, indeed, the use of the latter superplasticiser

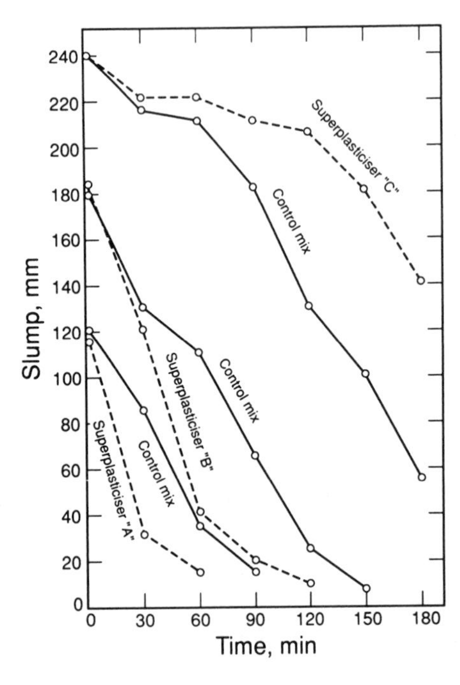

Fig. 4.8. Effect of superplasticisers on slump loss of concretes of different initial slumps. (Taken from the data of Ref. 4.17.)

considerably reduced the rate of slump loss and, consequently, the slump of the mix after 3 h remained comparatively high (i.e. 140 mm) and more than adequate for most concreting purposes. Anyway, superplasticisers, can, in general, be used successfully in hot weather conditions because they facilitate a considerable increase in the initial slump, and thereby overcome subsequent slump loss. In this respect it may be noted that sometimes superplasticisers are used, not only to increase the slump to the desired level but, simultaneously, to also reduce the amount of mixing water. In turn, this reduction can be utilised to reduce the cement content or, alternatively, to impart to the concrete improved properties due to the lower W/C ratio. Furthermore, under more severe conditions, where such an increase in the initial slump is not enough, superplasticisers may be used successfully for retempering. This specific subject is dealt with later in the text (see section 4.4.3.2).

4.3.3. Fly-Ash

Fly-ash, ground blast-furnace slag and pozzolans are used sometimes as a partial replacement of Portland cement (Chapter 3). In hot weather conditions this replacement may be deemed desirable because it reduces the rate of heat evolution, and thereby reduces the rise in concrete temperature and its associated adverse effects on concrete properties, including the rate of slump loss.

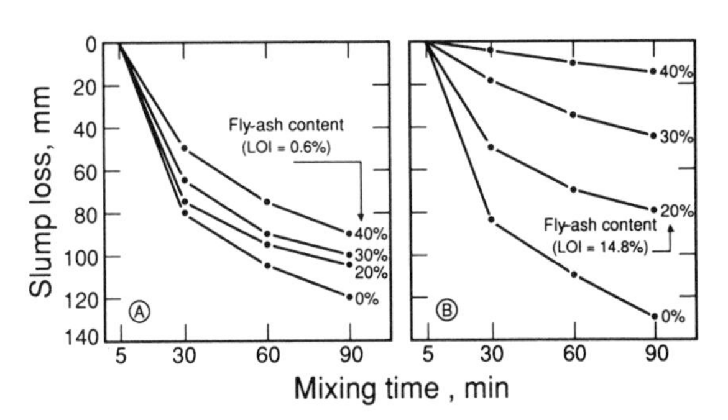

Fig. 4.9. Effect of replacing the cement with type F fly-ash (ASTM 618) on the rate of slump loss at 30°C. Loss of ignition of (A) fly-ash 0·6%, and of (B) fly-ash 14·8%. (Adapted from Ref. 4.20.)

Indeed, the replacement of the Portland cement by type F fly-ash (i.e. fly-ash originating from bituminous coal) was found to reduce the rate of slump loss in a prolonged mixed concrete, and this reduction increased with the increase in the percentage of the cement replaced (Fig. 4.9). This effect cannot be attributed only to the resulting lower cement content, and the associated lower heat of hydration, because it was found that replacing the cement by identical amounts of fine sand hardly affected slump loss. That is, the use of fly-ash as such, for reasons which are not clear as yet, brought about the reduction in the rate of slump loss.

The beneficial effect of fly-ash on the rate of the slump loss was found to be related to its loss on ignition (LOI), i.e. a higher LOI brought about a greater reduction in the rate of slump loss (Fig. 4.9). Again, it is rather difficult to explain this observation, and in no way is it to be regarded as a recommendation to use high LOI fly-ash in concrete. The latter may be desirable with respect to slump loss, but it must be remembered that a high LOI, which indicates the unburnt coal content in the ash, may be detrimental to the remaining properties of fly-ash concrete. Hence, regardless of the above finding, the use of fly-ash with a high LOI should be avoided.

4.3.4. Long Mixing and Delivery Times

Agitation of the concrete, while being transported by a truck mixer, is employed in order to delay setting and facilitate long hauling periods. The continuous agitation results in a grinding effect which, among other things, delays setting by breaking up the structure which is otherwise formed by the hydration products. This effect is also associated with the removal of some of

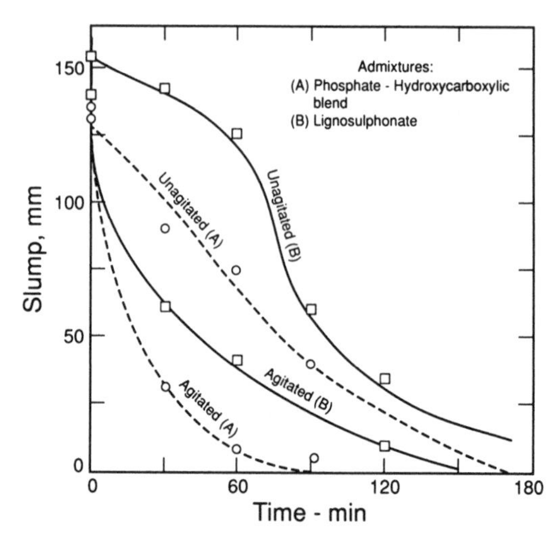

Fig. 4.10. Effect of continuous agitation on slump loss of concrete. (Adapted from Ref. 4.22.)

the hydration products from the surface of the hydrating cement grains, and thereby with the exposure of new surfaces to hydration. In other words, while setting is delayed due to breaking up of the structure, hydration is accelerated due to the greater exposure to water of the cement grains. A greater rate of hydration implies a greater rate of water consumption, and thereby a greater rate of slump loss. Moreover, the grinding effect produces fine material which increases the specific surface area of the solids in the mix. Consequently, more water is adsorbed and held on the surface of the solids, the amount of the free water in the mix is, thereby, reduced and rate of slump loss is further increased. In other words, it is to be expected that the rate of slump loss in a continuously agitated concrete will be greater than the corresponding rate in non-agitated concrete. This implication is reflected in the recommendations of the ACI Committee 305 [4.21] which state that 'the amount of mixing and agitating should be held to the minimum practicable', and 'consideration should be given to hauling concrete in a still drum instead of agitating on the way to the job'. This expected adverse effect of agitation on slump loss is confirmed by the data presented in Fig. 4.10 but not by the data presented in Fig. 4.11. In fact, the latter figure indicates that in plain concrete agitation slows, rather than accelerates, the rate of slump loss. In a retarded concrete, however, the slump loss is apparently independent of whether or not the concrete is agitated.

It may be also noted from Fig. 4.11 that the use of retarders increased considerably the slump loss of both agitated and non-agitated concrete. Accordingly, and considering the data discussed in section 4.3.2.3, the use of

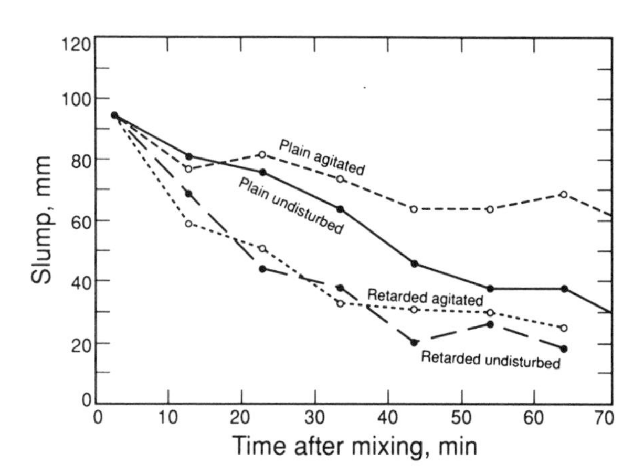

Fig. 4.11. Effect of continuous agitation on slump loss of concrete at 21–24°C. (Adapted from Ref. 4.14.)

retarders in agitated concrete may be questioned and, perhaps, even avoided altogether. Again, it should be pointed out that, in view of the considerable number of brands of admixtures available, the selection of the specific material to be used must be based on satisfactory past experience or on results of laboratory tests.

It is to be expected that longer delivery periods will be associated with a greater slump loss because of the longer hydration periods involved and the longer exposure time of the concrete to the grinding effect. Moreover, a further increase in the slump loss is to be expected with higher temperatures. These expected effects are confirmed by the data presented in Fig. 4.12 in which the amount of mixing water required to produce a slump of 100 mm, at the time of discharge, is plotted against the corresponding delivery time. In this presentation the greater water requirement implies a greater slump loss at the time of discharge. It can be seen that, indeed, slump loss increases with temperature and delivery time.

It may also be noted from Fig. 4.12 that the use of a water-reducing admixture or fly-ash (type F) was beneficial because it reduced the amount of mixing water which was required to control the slump at the time of discharge. It seems that in this respect fly-ash is preferable because its effect was less sensitive to delivery times.

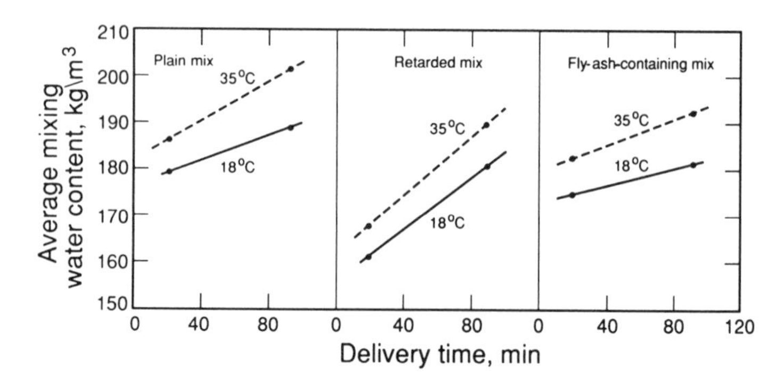

Fig. 4.12. Effect of delivery time and temperature on the amount of mixing water required to produce a 100 mm slump at the time of discharge. (Adapted from Ref. 4.23.)

4.4. CONTROL OF WORKABILITY

The consistency of the concrete mix, at the time of delivery, must be adequate to facilitate its easy handling without an appreciable risk of segregation. It is very important, therefore, to impart to the fresh concrete the required consistency, and in this respect the effect of elevated temperatures on slump loss must be considered and allowed for. The required slump depends on many factors such as the minimum dimensions of the concrete elements in question, the spacing of the reinforcing bars, etc. A minimum slump of 50 mm is sometimes quoted [4.22] which is also a typical truck mixer discharge limit. This value seems to be rather low for normal applications and a higher value, namely 75–100 mm, should be preferably considered, at least in the mix design stage [4.21]. The time after mixing when the desired slump is required may vary considerably. It may be 30 min or less when the concrete is produced *in situ* and 2–3 h and, even more, when long distance hauling is involved. Of course, the longer the hauling time and the higher the ambient temperature, the more difficult it is to overcome slump loss and to give the concrete the desired consistency at the time of discharge.

In principle, the accelerating effect of high temperatures on slump loss may be overcome by using one, or some, of the following methods which are schematically described in Fig. 4.13.

(1) Using a wetter mix, that is a mix with a higher initial slump. The rate of slump loss in high slump mixes is known to be higher than the rate in low slump mixes. However, if the initial slump is high enough, the

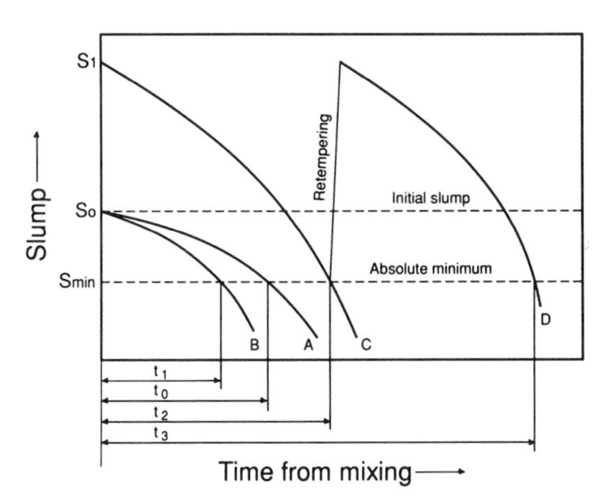

Fig. 4.13. Schematic representation of possible methods to overcome the effect of high temperatures on slump loss.

residual slump may remain higher than the slump required when the concrete is used. The higher slump can be produced either by using an increased amount of mixing water or by the use of water-reducing admixtures.

(2) Reducing the initial concrete temperature either by keeping it as close as possible to ambient temperatures, or by lowering it below this level, mainly by the use of cold water or ice.

(3) Retempering of the mix, i.e. restoring the initial slump of the fresh concrete by remixing with additional water or a suitable superplasticiser.

Curve A in Fig. 4.13 represents the slump loss with time in a concrete mix subjected to moderate temperatures. Having the initial slump, S_0, it reaches the absolute minimum, S_{min}, at the time t_0 after mixing, when in this context the absolute minimum is the lowest slump which allows the concrete to be properly handled and compacted. When the same mix is subjected to higher temperatures, the rate of slump loss is increased and the absolute minimum is reached after a shorter time, t_1, which may be not long enough under conditions considered (curve B). In order to extend the workable time of the mix, the initial slump may be increased to S_1. The rate of slump loss of this high slump mix (curve C) is greater than the mix having the initial slump, S_0, but, nevertheless, the mix remains workable for the longer time, t_2. If the time, t_2, is not long enough, the workable time of the mix can be further extended to t_3 by retempering (curve D). Finally, by lowering concrete temperature, the rate of slump loss is reduced, and may be represented by curve A instead of curve B.

The efficiency of the above-mentioned methods is reflected, to some extent, in the schematic representation of Fig. 4.13. It may be noted that, generally, lowering the concrete temperature may constitute a solution when relatively short workable times are required. Using a wetter mix may result in somewhat longer times and retempering in the longest ones.

4.4.1. Increasing Initial Slump

The most obvious and convenient way to increase initial slump is by increasing the amount of mixing water. In practice, water may be used to produce slumps not higher than, say, 150–180 mm, because wetter mixes usually exhibit an excessive tendency to segregate. A further limitation of the increase in the amount of water involves its effect on the W/C ratio, and thereby on concrete properties. This effect, however, can be avoided by a corresponding increase in the cement content to allow for the increased W/C ratio.

An increased cement content is not necessarily desirable because it gives the concrete a higher drying shrinkage and as such makes it more susceptible to cracking. It is preferable, therefore, to use water-reducing admixtures, either conventional or high range (superplasticisers), instead of water, in order to increase the initial slump of the mix. That is, in this application the admixtures are not used to reduce the amount of mixing water but to increase the fluidity of the mix. This may be somewhat different when superplasticisers are used. The latter, being much more effective water reducers, may sometimes allow the simultaneous reduction in the amount of mixing water and the increase in slump.

4.4.2. Lowering Concrete Temperature

In this section the lowering of concrete temperature is discussed mainly with respect to accelerated slump loss which is brought about by high ambient temperatures. This is, of course, a very important aspect and warrants by itself an adequate and satisfactory solution. Nevertheless, keeping the concrete temperature as low as possible is also highly desirable in order to reduce the adverse effect of the higher temperatures on concrete strength, its vulnerability to thermal cracking, etc. Accordingly, it may be argued that lowering concrete temperature is to be preferred to increasing its initial slump in order to counteract the accelerated slump loss. This may be the case, but the cooling operation is costly and is usually economically feasible only in big projects where large quantities of concrete are produced and placed.

A few means are available to keep concrete temperature as low as possible, and most of them are self-evident. Insulating water supply lines and tanks, shading of materials and concrete-making facilities from direct sunshine, and sprinkling the aggregates with clean uncontaminated water, for example, are such means. Other means include painting the drums of truck mixers and cement silos white to reduce heat gain. The use of hot cement should be avoided, although the relatively high temperature of 77°C is sometimes quoted as the maximum limit [4.21]. It may be noted all these means limit the heat gain of the concrete and its ingredients, and therefore may keep concrete temperature, at best, not too much higher than ambient temperatures. Consequently, these means are mostly used in conjunction with other means which are capable of lowering concrete temperature below ambient temperatures. These include the use of cooled materials, and in particular, the use of cold water or crushed ice.

4.4.2.1. Use of Cold Water

The initial concrete temperature which is brought about by the use of cold water, can be estimated from the following heat equilibrium equation on the assumption that the specific heat of the solids in the mix is the same and equals 0·22:

$$T_{conc} = \frac{0{\cdot}22(T_a W_a + T_c W_c) + T_w W_w}{0{\cdot}22(W_a + W_c) + W_w} \tag{4.1}$$

where T_{conc}, T_a, T_c and T_w are the temperatures (°C) of the concrete, aggregate, cement and water, respectively; and W_a, W_c and W_w are the weights (kg) of the aggregate, cement and water, respectively.

Substituting $a = W_a/W_c$ (i.e. aggregate to cement ratio) and $\omega = W_w/W_c$ (i.e. water to cement ratio), and assuming that the specific heat of the solids is 0·2, the above equation takes a somewhat simplified form:

$$T_{conc} = \frac{T_c + aT_a + 5\omega T_w}{1 + a + 5\omega} \tag{4.2}$$

In practice, water can be cooled down to, say, 5°C. Considering an ordinary mix where $a = 6$ and $\omega = 0{\cdot}6$, the estimated concrete temperatures are 22·5,

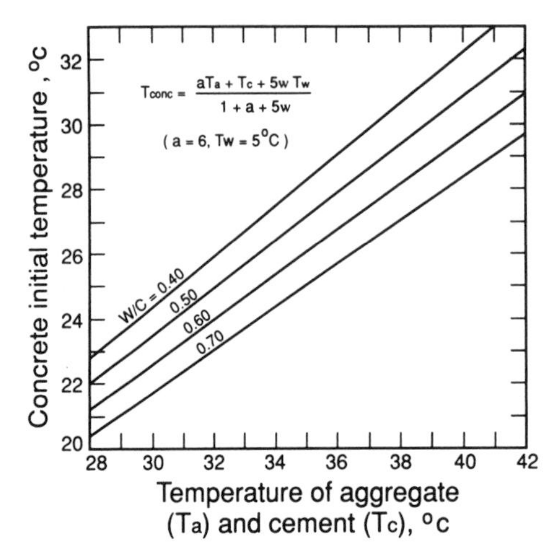

Fig. 4.14. Graphical solution of eqn (4.2) for aggregate to cement ratio $a = 0{\cdot}6$ and mixing water temperature $T_w = 5$°C.

26 and 29·5°C when the cement and aggregate temperatures are 30, 35 and 40°C, respectively (Fig. 4.14). Alternatively, in order to lower concrete temperature by 1°C, the wa‹er temperature has to be lowered by 3·3°C. Hence, it may be concluded that use of cold water can reduce concrete temperature by up to ~10°C. In practice, however, this is not the case and the maximum reduction in concrete temperature that can be obtained by using cold water is, apparently, about 6°C [4.21].

The cooling of water may be achieved by mechanical refrigeration, the use of crushed ice and also by injecting liquid nitrogen into the water tank. Such means, although costly, can produce only a moderate reduction in concrete temperature, i.e. as mentioned previously, a maximum reduction of about 6°C. In fact even a lower maximum of 3–5°C is sometimes mentioned [4.24].

4.4.2.2. Use of Ice

A further reduction in the initial temperature of the fresh mix can be achieved by using ice as part of the mixing water. The ice is introduced into the mix in the form of crushed, chipped or shaved ice, and on melting during the mixing operation absorbs heat at a rate of 79·6 kcal/kg (335 J/g), and thereby lowers the temperature of the concrete. Assuming the ice temperature is 0°C, and using the same notation as in eqn (4.1), the estimated concrete temperature is given by (W_i is the weight of the ice):

$$T_{conc} = \frac{0{\cdot}22(T_a W_a + T_c W_c) + T_w W_w - 79{\cdot}6 W_i}{0{\cdot}22(W_a + W_c) + W_w + W_i} \tag{4.3}$$

Substituting $a = W_a/W_c$, $\omega = (W_i + W_w)/W_c$ and $\alpha = W_i/(W_i + W_w)$,

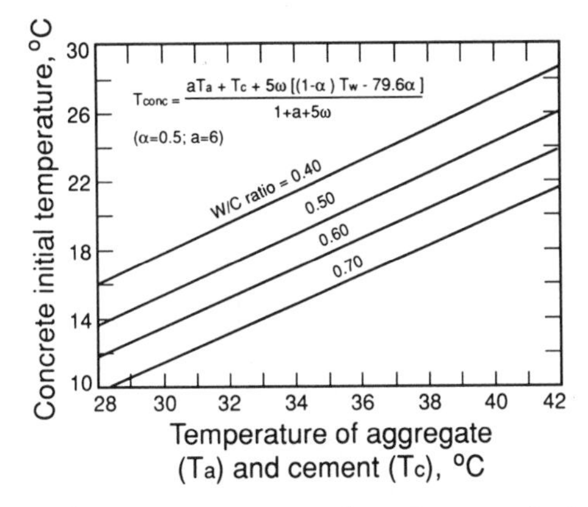

Fig. 4.15. Graphical solution of eqn (4.4) for aggregate to cement ratio a = 0·6, and substituting ice for 50% of total mixing water (α = 0·50).

and assuming, again, that the specific heat of the solids is 0·2, eqn (4.3) takes the following simplified form:

$$T_{\text{conc}} = \frac{T_c + aT_a + 5\omega(T_w(1 - \alpha) - 79{\cdot}6\alpha)}{a + 1 + 5\omega} \tag{4.4}$$

In order to facilitate a more rapid mixing of concrete ingredients, some part of the mixing water, usually not less than 25%, is added as liquid water. That is, the amount of water which is added in the form of ice usually does not exceed 75% of the total. Considering a more moderate value of 50%, and the mix previously investigated (i.e. a = 6 and ω = 0·6), it is found, by solving eqn (4.3) or eqn (4.4), that the estimated concrete temperature for $T_a = T_c = T_w$ = 30, 35 and 40°C is 13·5, 17·6 and 22°C, respectively (Fig. 4.15). That is, under the conditions considered, the use of ice may reduce concrete temperature by up to 18°C, and a higher reduction may be achieved if a greater part of the mixing water (i.e. 75%) is introduced into the mix in the form of ice. Again, apparently in practice such a considerable reduction cannot be achieved, and the maximum obtainable to be considered is about 11°C [4.21].

The use of ice is conditional on the availability of a suitable and reliable source of ice. When block ice is supplied, refrigerated storage must be provided as well as suitable mechanical means to crush the ice. The need for such means can be avoided if the ice is produced on site in the form of flakes. Again, using ice to cool the concrete is a costly procedure and may be economic only under specific conditions.

4.4.2.3. Use of Cooled Aggregate

The coarse aggregate constitutes some 50% of concrete ingredients and it is to be expected, therefore, that the use of cooled coarse aggregate will bring about a considerable reduction in concrete temperature. Again, this effect can be estimated quantitatively by solving eqn (4.1). Considering a typical mix in which $W_a = 1800$ kg (of which 1200 kg is coarse aggregate and the remaining 600 kg is fine aggregate), $W_c = 330$ kg and $W_w = 200$ kg, the estimated concrete temperature for $T_c = T_w = 30°C$ and $T_a = 20°C$ for the coarse aggregate only, will be some 26°C. That is, in order to reduce concrete temperature by 1°C, the temperature of the coarse aggregate must be reduced by 2·5°C.

One way to cool the aggregate is by using cold water for spraying or inundating. This procedure requires, of course, great quantities of clean and uncontaminated water which are not always available in hot arid areas. Wetting the aggregate involves the presence of free moisture which must be allowed for by an appropriate reduction in the amount of water which is added to the mix. Blowing air through the wet aggregate, due to the increased evaporation, will bring about a greater reduction in aggregate temperature. If cold air is used, a further reduction may be achieved, and the temperature of the aggregate may go down as low as 7°C [4.21]. Cooling by air is, again, a costly operation which may be justified only under specific conditions.

Another method for cooling of the coarse aggregate involves the use of liquid nitrogen. In this method the aggregate is sprayed upon liquid nitrogen and the resulting cold gas is drawn through the aggregate by a fan [4.25, 4.26]. It is claimed that by using this method, the temperature of a dry aggregate can be brought down to −18°C [4.24].

It may be pointed out that liquid nitrogen may also be used to lower concrete temperature by injecting it directly into the fresh mix. This method has been reported to be effective in lowering concrete temperature without adversely affecting its properties.

4.4.3. Retempering

Retempering is defined as 'addition of water and remixing of concrete or mortar which has lost enough workability to become unplaceable or unstable' [4.27]. In practice, however, a wider definition is usually adopted, to include later additions of superplasticisers as well. Restoring the required slump (workability) of the concrete mix by retempering is particularly useful when long hauling periods and extreme weather conditions are involved, whereas the use of wet mixes with a high initial slump, is suitable for short delivery periods and moderate weather conditions.

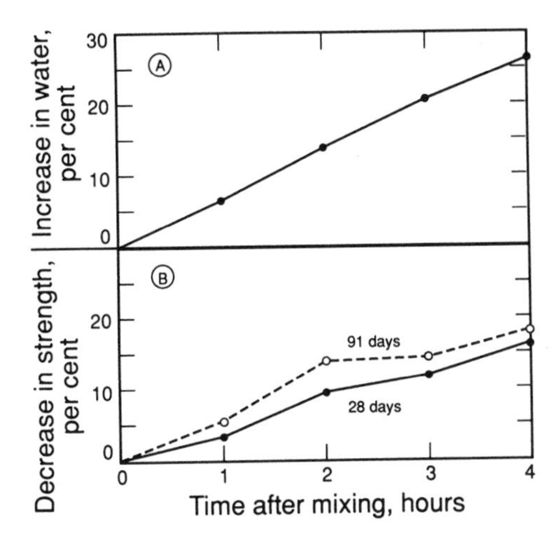

Fig. 4.16. Effect of time elapsed after mixing on (A) the increase in the amount of water required for retempering to the initial slump of 75 mm and, (B) the resulting decrease in compressive strength. (Taken from the data in Ref. 4.28.)

4.4.3.1. Retempering with Water

In this method, concrete is prepared with the required slump and is later retempered with an amount of water which is just sufficient to restore the slump to its initial level. Concrete properties and, indeed, its quality in general, are determined under otherwise the same conditions, by the W/C ratio. The addition of water for retempering increases this ratio, and thereby concrete strength, for example, is adversely affected. This expected effect is demonstrated in Fig. 4.16 for concrete mixed and retempered in the ambient temperature range of 25–38°C (concrete temperature 25–33°C). It may also be noted that the amount of water required for retempering increased with the increase in time after mixing.

The adverse effect of the retempering water on concrete properties may be overcome in two ways. In the first one, the water is simply added together with a corresponding amount of cement which is required to keep the W/C ratio unchanged. In the second, the additional amount of the retempering water is allowed for in the selection of mix proportions, and the cement content is determined, in the first instance, so that when the retempering water is added, the required W/C ratio is not exceeded. This is not always easy to achieve because a fair estimate of the amount of retempering water, which will be subsequently needed, must be known at the mix design stage.

It may be noted that both ways of offsetting the adverse effect of the retempering water on concrete properties involve increased cement content. This may be deemed undesirable because of the associated increase in heat evolution which further aggravates the problem, and also because the higher cement content increases shrinkage, and thereby the risk of shrinkage cracking. The use of conventional admixtures (i.e. types A and D) or superplasticisers

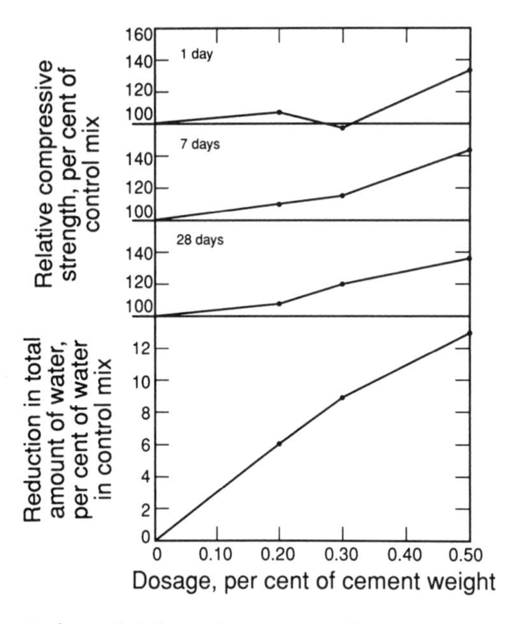

Fig. 4.17. Effect of admixture type A on the reduction in total amount of mixing water and the resulting compressive strength. Basic slump 100 mm, temperature 30°C. Retempering 1 h after mixing. (Taken from the data of Ref. 4.10.)

is beneficial in this respect because it does not involve an increased cement content. Moreover, the strength of the concrete may be favourably affected, in particular when greater dosages than the recommended ones are used. This beneficial effect of admixtures is reflected in the data presented in Fig. 4.17.

The total amount of mixing water in Fig. 4.17 is the combined amount of water required to produce the initial slump of 100 mm and the amount required subsequently for retempering in order to restore the slump to its initial level. It can be seen that the use of the water-reducing admixture resulted in a reduction in this total amount of water, and this reduction increased with the increase in the amount of admixture used.

The reduction in the total amount of water, lowers the corresponding W/C ratio, and strength is expected, therefore, to increase. This is indeed the case as may be noted from Fig. 4.17. It must be realised, however, that when water-reducing admixtures are used, the amount of water required for retempering is not less than the amount required when no such admixtures are used. In fact, in both cases virtually the same amount is needed, and the reduced amount of the total is due to the reduced amount which is needed to give the mix the initial slump. This may be concluded from the data presented in Fig. 4.18 which relate to mixes with the same cement content and the same initial slump of 90 mm, which were retempered 2 h after mixing.

It may be noted from Fig. 4.18 that retempering increased the W/C ratio by 0·06 in all mixes, the one exception being the mix containing the superplasticiser, in which the increase was slightly greater, i.e. 0·07. The cement content

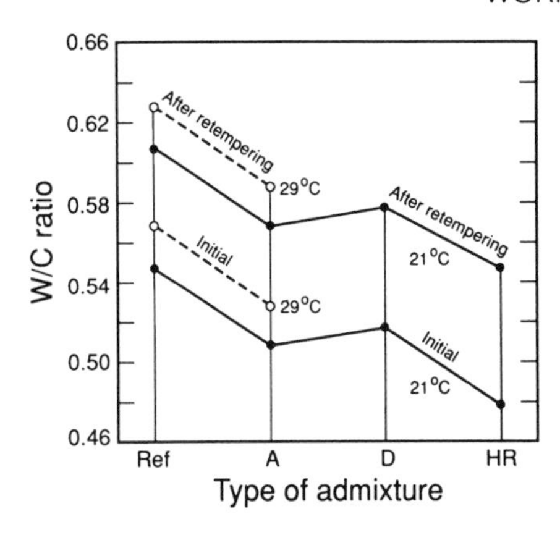

Fig. 4.18. Effect of water reducing admixtures on the W/C ratio of retempered mixes. Initial slump 90 mm, retempering 2 h after mixing. (Taken from the data of Ref. 4.8.)

in all mixes being the same, the same increase in the W/C ratio implies that the same amount of water was used for retempering in all mixes. As the initial W/C ratio of the admixture-containing mixes was lower than the W/C ratio of the reference mix, on the one hand, and the increase in the W/C ratio on retempering was the same, on the other, the W/C ratio of the admixture-containing mixes remained lower. Hence, in agreement with the data of Fig. 4.17, it is to be expected that the latter mixes will exhibit a higher strength than the reference mix.

It was pointed out earlier (see section 4.3.2.2) that water-reducing admixtures usually accelerate the rate of slump loss. Nevertheless, the use of such admixtures in retempered mixes should be favourably considered, because such use does not involve either a reduced strength nor a higher cement content. Moreover, when higher temperatures are considered, the use of increased dosages of water-reducing admixtures may provide a practical solution to the increased amount of water needed for retempering.

4.4.3.2. Retempering with Superplasticisers

Superplasticisers considerably increase the fluidity of the fresh concrete and as such may be used, and indeed are used, for retempering. In most cases superplasticisers increase the rate of slump loss (see section 4.3.2.4) but, on the other hand, their use increases neither the W/C ratio nor the cement content. Superplasticisers can be used for retempering of both plain or superplasticised concrete, i.e. for retempering of concrete in which no superplasticisers were added initially as well as for retempering of concrete in which superplasticisers were added to the original mix. In the latter case the superplasticiser may be utilised to reduce the amount of mixing water, or the cement content, or both.

It was shown earlier (Fig. 4.7) that slump loss is increased with temperature.

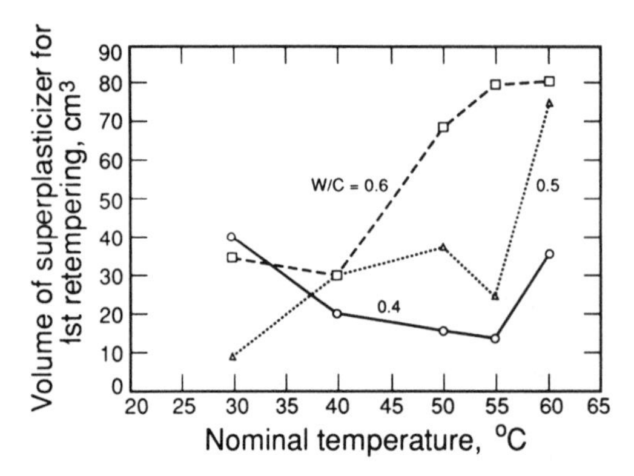

Fig. 4.19. Effect of temperature on dosage of superplasticiser required to restore slump on retempering to the initial level of 100 mm. (Adapted from Ref. 4.29.)

It is to be expected, therefore, that the quantity (dosage) of superplasticiser required for retempering (i.e. to restore the slump to its initial level) will also increase with temperature. However, experimental data relating to this aspect are not always clear and can only partly be explained. The data presented in Fig. 4.19, for example, indicate that the effect of temperature on the amount of superplasticiser required for retempering depends on the W/C ratio of the concrete involved. That is, the quantity of the superplasticiser required for retempering remained virtually the same, and unrelated to temperature, when the W/C ratio was 0·4. When the W/C ratio was 0·5, it became temperature dependent mostly at the high temperature level of 55–60°C, and only at the high W/C ratio of 0·6 did it become temperature dependent, and increase with the latter, in the wider range of 40–60°C. On the other hand, the data presented in Fig. 4.20 show the reverse trend, namely, that the required dosage

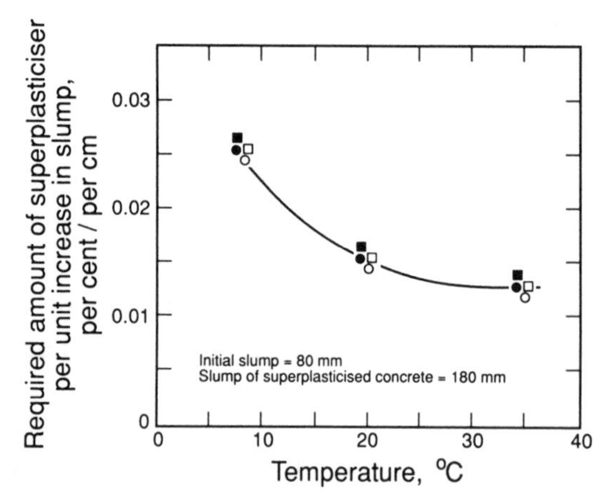

Fig. 4.20. Effect of temperature on dosage of superplasticiser required to increase slump from 80 to 180 mm. (Adapted from Ref. 4.6.)

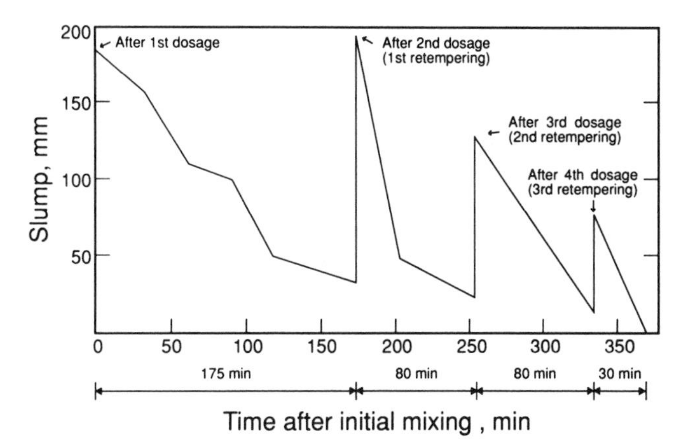

Fig. 4.21. Effect of repeated retempering with superplasticiser on concrete slump. (Adapted from Ref. 4.30.)

actually decreases with temperature. This trend, however, although quite evident in the lower temperature range of 7–20°C, is hardly apparent in the higher range of 20–30°C. Noting that the data of Fig. 4.19 relate to a much higher temperature range of 30–60°C, it may be argued that the data of the two figures in question are not comparable and, therefore, not necessarily contradictory. As mentioned earlier the exact nature of the effect of temperature on the dosage required for retempering is not clear. Nevertheless, it is usually assumed that a greater dosage of superplasticiser is required under higher temperatures [4.29].

Concrete may be retempered more than once. The efficiency of the superplasticiser, however, diminishes as the number of retemperings is increased. That is, a lower slump is reached, and accordingly the length of time in which the concrete remains workable becomes shorter, if the same dosage is repeated in the successive retemperings (Fig. 4.21). This observation is further supported by the data presented in Fig. 4.22(A). This part of the figure presents the effect of the time elapsed, from the initial mixing to retempering, on slump loss in a concrete retempered with 0·5% of a superplasticiser. In the case considered, retempering was carried out after 30, 60 and 90 min, and it can be seen that, in agreement with the data of Fig. 4.21, the effect of the 0·5% dosage decreased with the increase in the time of retempering. Such a decrease, however, was not observed when a high dosage of 3% was used and, in fact, the higher dosage was also more effective in increasing concrete slump (Fig. 4.22(B)). Hence, it may be concluded that a higher dosage of superplasticiser can be used efficiently to counteract the diminishing effect of time on the effectiveness of retempering. Such an increase of dosage must be exercised,

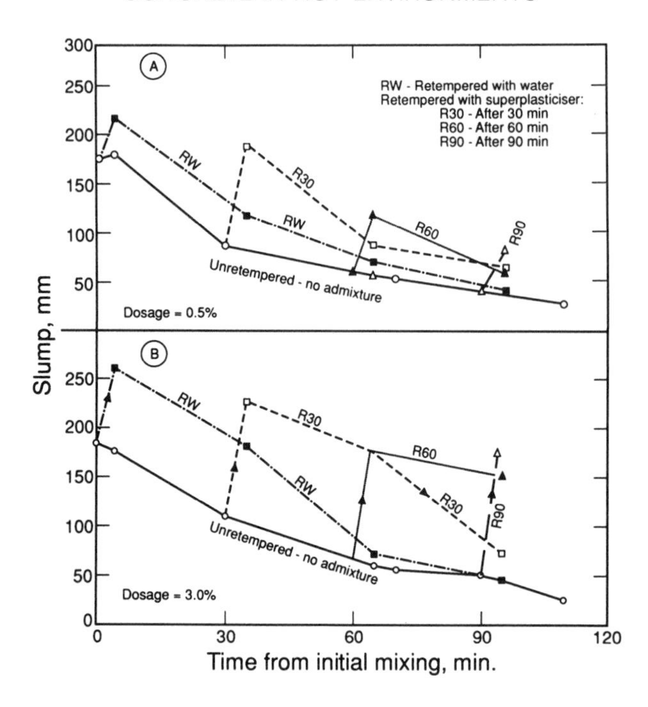

Fig. 4.22. Effect of retempering at different times with different dosages of superplasticiser on slump loss. (Adapted from Ref. 4.31.)

however, with due care because at a certain level it may give the fresh concrete an excessive tendency to segregation.

It was pointed out earlier that the use of superplasticisers for retempering does not involve an increase in the W/C ratio, and in some cases may even facilitate a reduction in the latter ratio. Hence, considering the possible effect of the W/C ratio alone, it is to be expected that the properties of a retempered concrete will be essentially the same as the properties of otherwise the same unretempered concrete. On the other hand, retempering as such hinders structure formation and, thereby, may adversely affect such properties as strength, etc. Data relating to the effect of retempering on compressive strength are presented in Fig. 4.23. It can be seen that, indeed, retempering adversely affected the earlier strength at 7 days. This effect, however, was rather small (i.e. a reduction of some 5% for twice retempered concrete), and virtually disappeared at the age of 28 days. Tests relating to flexural strength, splitting tensile strength, static modulus of elasticity and pulse velocity lead to the same conclusion, i.e. that retempering does not affect significantly the properties of the hardened concrete [4.29, 4.30].

In passing it must be stressed again that admixtures are commercially

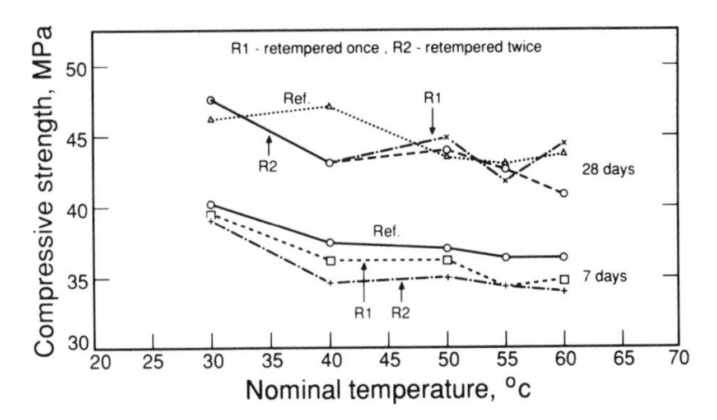

Fig. 4.23. Effect of repeated retempering with a superplasticiser on compressive strength of concrete with a W/C ratio of 0·40. R1—retempered once, R2—retempered twice. (Adapted from Ref. 4.29.)

produced and although complying with the very same standards, may differ considerably in their composition and properties. Hence, due caution should be exercised in adopting the preceding conclusions in practical applications. The selection of admixtures for a specific use must be based, always, on past experience or on tests data relevant to the intended use.

4.5. SUMMARY AND CONCLUDING REMARKS

Workability is 'the property determining the effort required to manipulate a freshly mixed quantity of concrete with a minimum loss of homogeneity' (ASTM C115). It is determined by the consistency and the cohesiveness of the mix, but once cohesiveness is attained by proper selection of materials and mix proportions, workability is further controlled by the consistency alone. Consistency, in turn, is controlled by the amount of mixing water and the use of admixtures, and is determined quantitatively by the slump or the Vebe tests or by the compacting apparatus. With time, due mainly to the hydration of the cement, the concrete stiffens and its slump decreases. Hence, reference is made to 'slump loss'. The rate of slump loss increases with temperature because of the accelerating effect of temperature on the rate of hydration and the rate of evaporation. The increased rate of stiffening, brought about by elevated temperatures, constitutes a serious problem under hot weather conditions, and is further aggravated when long hauling periods are involved. In order to allow for the increased loss of slump, wet mixes, with a high initial slump of

180–200 mm, are sometimes used. The higher slump is produced either by increasing the amount of mixing water or by using water-reducing admixtures (conventional or high range). The use of type F fly-ash may be beneficial in this respect. Other means include the lowering of concrete temperature by using cold water, the substitution of ice for part of the mixing water, and sometimes, also, the use of cooled coarse aggregate. Retempering, i.e. remixing with additional water or superplasticisers, is mostly used when long hauling periods and extreme weather conditions are involved.

REFERENCES

4.1. ACI Committee 211, Standard practice for selecting proportions for normal and heavyweight and mass concrete (ACI 211.1–89). In *ACI Manual of Concrete Practice* (Part 1). ACI, Detroit, MI, USA, 1990.
4.2. Teychenne, D.C., Franklin, R.E. & Erntroy, H.C., *Design of Normal Concrete Mixes* (reviewed edn). Dept. of the Environment, Building Research Establishment, Garston, Watford, UK, 1988.
4.3. Popovics, S., Relation between the change in water content and the consistency of fresh concrete. *Mag. Concrete Res.*, **14**(4) (1962), 99–108.
4.4. Klieger, P., Effect of mixing and curing temperature on concrete strength. *Proc. J. ACI*, **54**(12) (1958), 1063–81.
4.5. US Bureau of Reclamation, *Concrete Manual* (8th edn, revised). Denver, CO, USA, 1981, Fig. 118, p. 256.
4.6. Yamamoto, Y. & Kobayashi, S., Effect of temperature on the properties of superplasticized concrete. *Proc. ACI*, **83**(1) (1986), 80–6.
4.7. Mahter, B., The warmer the concrete the faster the cement hydrates. *Concrete Int.*, **9**(8) (1987), 29–33.
4.8. Previte, R.W., Concrete slump loss. *Proc. J. ACI*, **74**(8) (1977), 361–7.
4.9. Hampton, J.S., Extended workability of concrete containing high-range water-reducing admixtures in hot weather. In *Development in the Use of Superplasticizers* (ACI Spec. Publ. SP 68). ACI, Detroit, MI, USA, 1981, pp. 409–22.
4.10. Ravina, D., Retempering of prolonged mixed concrete with admixtures in hot weather. *J. ACI*, **72**(6) (1975), 291–5.
4.11. Meyer, L.M. & Perenchio, W.F., Theory of concrete slump loss as related to the use of chemical admixtures. *Concrete Int.*, **1**(1) (1979), 36–43.
4.12. Hersey, A.T., Slump loss caused by admixtures. *Proc. ACI*, **74**(10) (1975), 526–7.
4.13. Perenchio, W.F., Whiting, D.A. & Kantro, D.L., Water reduction, slump loss,

and entrained air–void systems as influenced by superplasticizer. In *Superplasticizers in Concrete*. (ACI Spec. Publ. SP 68). ACI, Detroit, MI, USA, 1979, pp. 137–55.

4.14. Tuthill, L.H., Adams, R.F. & Hemme, J.M., Jr, Observation in testing and use of water-reducing retarders. In *Symp. on Effect of Water-Reducing Admixtures and Set Retarding Admixtures on Properties of Concrete* (ASTM Spec. Tech. Publ. No. 266). ASTM, Philadelphia, PA, USA, 1959, pp. 107–17.

4.15. Ramachandran, V.S., Feldman, R.F. & Beaudoin, J.J., *Concrete Science*. Heyden & Sons Ltd, Philadelphia, PA, USA, 1981, pp. 137–8.

4.16. Mailvaganam, N.P., Factors influencing slump loss in flowing concrete. In *Superplasticizers in Concrete*. (ACI Spec. Publ. SP 62). ACI, Detroit, MI, USA, 1979, pp. 389–403.

4.17. Collepardi, M., Guella, M.S. & Maniscalco, V., *Superplasticized Concrete in Hot Climates*. Giorante AICAP, Bari, Italy, 1983.

4.18. Gulyas, R.J., Hot weather concreting: Some problems and solutions. *Concrete Products*, **Aug**. (1988), 22–3.

4.19. Shilstone, J.M., Concrete strength loss and slump loss in summer. *Concrete Construct*., **May** (1982), 429–32.

4.20. Ravina, D., Slump loss of fly ash concrete. *Concrete Int*., **6**(4) (1984) 35–9.

4.21. ACI Committee 305, Hot-weather concreting (ACI 305R-89). In *ACI Manual of Concrete Practice* (Part 2). ACI, Detroit, MI, USA, 1990.

4.22. McCarthy, M., Tests on set retarding admixtures. *Precast Concrete*, **10**(3) (1979) 128–30.

4.23. Gaynor, R.D., Meininger, R.C. & Khan, T.S., Effect of temperature and delivery time on concrete proportions. In *Temperature Effects on Concrete* (ASTM Spec. Tech. Publ., STP 858). ASTM, Philadelphia, PA, USA, 1985, pp. 66–87.

4.24. Tipler, T.J., Handling. In *Proc. Intern. Seminar on Concrete in Hot Countries*. Helsingor, 1981, Skanska, Malmo, Sweden, pp. 71–9.

4.25. Anon., Keeping it cool with liquid nitrogen. *Concrete Construct*., **25**(8) (1980), 606, 609.

4.26. Anon, Cooling concrete mixes with liquid nitrogen. *Concrete Construct*., **22**(5) (1977), 257–8.

4.27. ACI Committee 116, Cement and concrete terminology (ACI 116R-85). In *ACI Manual of Concrete Practice* (Part 1). ACI, Detroit, MI, USA, 1990.

4.28. Adams, R.F., Stodola, P.S. & Mitchel, D.R., Discussion of Ref. 4.26, *Proc. ACI*, **59**(9) (1962), 1249–50.

4.29. Samarai, M.A., Ramakrishnan, V. & Malhotra, V.M., *Effect of Retempering with Superplasticizers on Properties of Fresh and Hardened Concrete Mixed at Higher Ambient Temperatures* (ACI Spec. Publ. SP 119). ACI, Detroit, MI, USA, 1989, pp. 273–95.

4.30. Ramakrishnan, V., Coyle, W.V. & Pande, S.S., Workability and strength of retempered superplasticized concretes. In *Superplasticizers in Concrete* (Transportation Res. Rec. TRR 720). National Research Board, Washington DC, 1979, pp. 13–18.

4.31. Ravina, D. & Mor, A., Effects of superplasticizers. *Concrete Int.*, **8**(7) (1986), 53–5.

Chapter 5

Early Volume Changes and Cracking

5.1. INTRODUCTION

Cracking of concrete may occur before hardening, i.e. when the concrete reaches the stage in which it is not plastic any more and, therefore, cannot accommodate early volume changes. Accordingly, the resulting cracks are known as 'pre-hardening cracks' or 'plastic cracks'. Generally, pre-hardening cracks, if occurring, develop a few hours after the concrete has been placed and finished. The mechanisms involved may be different and, accordingly, distinction is made between 'plastic shrinkage cracks' and 'plastic settlement cracks'.

5.2. PLASTIC SHRINKAGE

When the fresh concrete is allowed to dry contraction takes place. This contraction in the pre-hardening stage is known as 'plastic shrinkage', and is to be distinguished from shrinkage in the hardened stage which is known as 'drying shrinkage' (see Chapter 7). Plastic shrinkage may cause cracking during the first few hours after the concrete has been placed, usually at the stage when its surface becomes dry. Such cracks are characterised by a random map pattern (Fig. 5.1(A)) but sometimes they develop as diagonal cracks at approximately 45° to the edges of the slab (Fig. 5.1(B)). At other times the cracks may develop along the reinforcement, particularly when the reinforcement is close to the

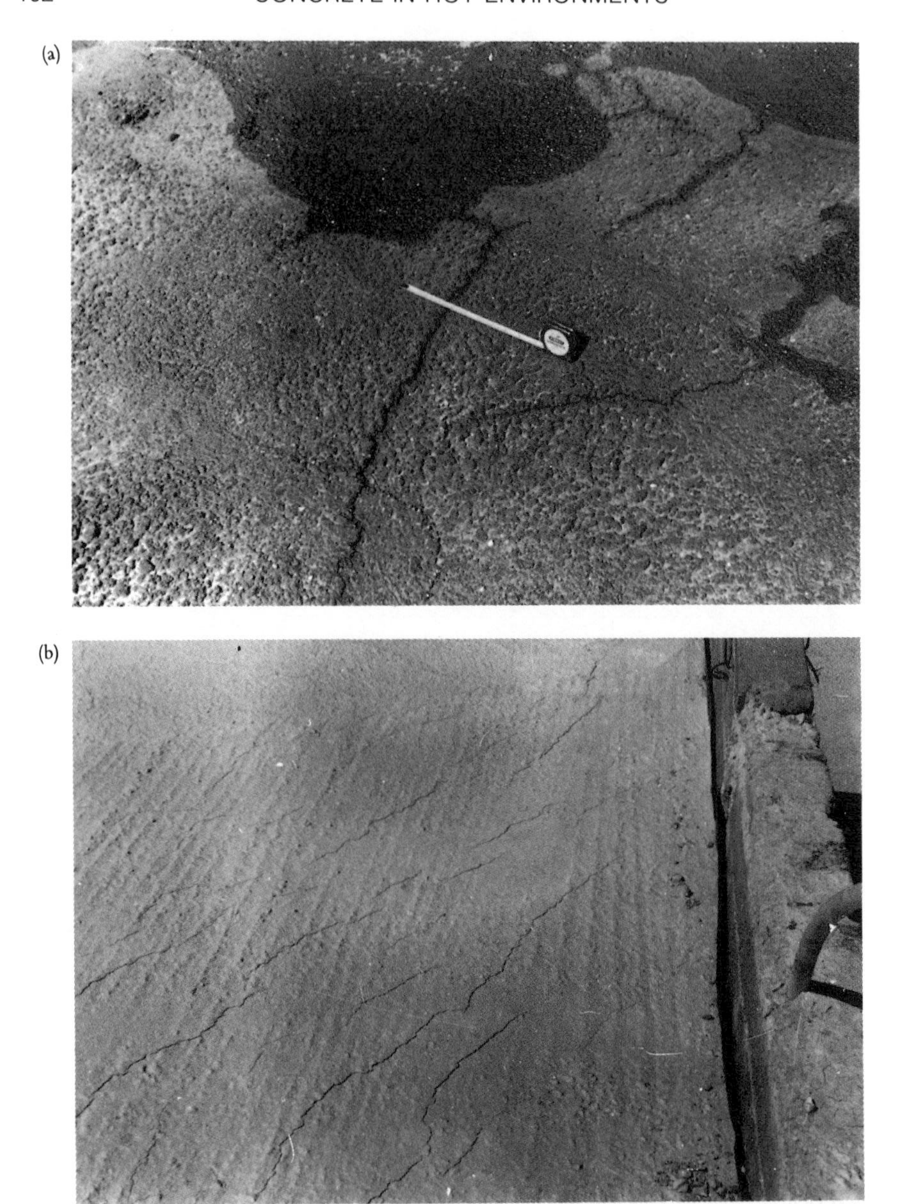

Fig. 5.1. Typical plastic cracking in a concrete slab.

surface. The width of the cracks varies and may reach a few millimeters. Similarly, their length varies from a few millimeters to 1 m and more. Usually, the cracks taper rapidly from the top surface, but, in extreme cases, a crack may penetrate the full depth of the slab.

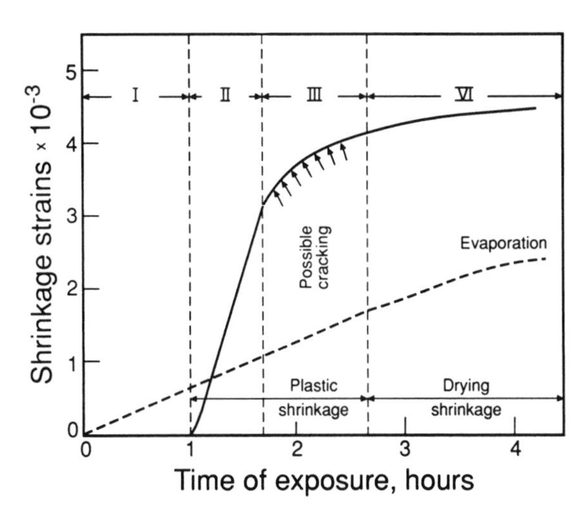

Fig. 5.2. Schematic description of early age shrinkage of concrete with time. (Adapted from Ref. 5.1.)

The drying, and the associated plastic shrinkage of fresh concrete, is schematically described in Fig. 5.2. Four stages are distinguishable.

Stage I —Rate of bleeding is greater than the rate of drying. Consequently, the surface of the concrete remains wet and no shrinkage takes place.

Stage II —Rate of drying is greater than the rate of bleeding. The surface dries out and shrinkage starts to take place. No cracking occurs because the concrete is still plastic enough to accommodate the resulting volume changes. Drying, and the corresponding shrinkage, proceed roughly at a constant rate.

Stage III—Concrete becomes brittle; restraint of shrinkage induces tensile stresses in the concrete which cracks, if and when its tensile strength is lower than the induced tensile stresses.

Stage IV—Concrete is set and drying shrinkage begins.

It was pointed out earlier that early drying of the fresh concrete results in plastic shrinkage which may cause cracking if and when the induced tensile stresses exceed the tensile strength of the concrete at the time considered. It still has to be explained why the drying of the concrete, as such, brings about plastic shrinkage. It has been suggested that the mechanism involved is that of capillary tension which, in turn, induces compressive stresses in the fresh concrete, and thereby causes its contraction, i.e. its plastic shrinkage [5.2]. A more detailed discussion of the mechanism of capillary tension is presented

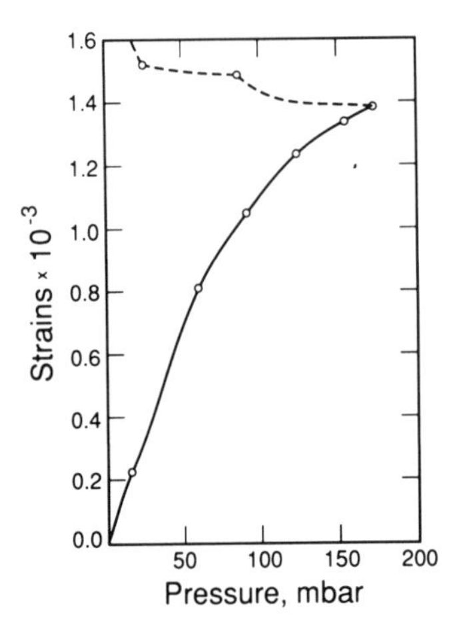

Fig. 5.3. The relation between plastic shrinkage and capillary pressure (Adapted from Ref. 5.2.)

later in this book (section 7.3.1), but it can be shown that this mechanism becomes operative when menisci are formed between the solid particles in the concrete surface. At the initial stage the concrete is still plastic and can be consolidated by the resulting pressure. Hence, plastic shrinkage occurs. This suggested mechanism is compatible with the observation that plastic shrinkage begins when the concrete surface becomes dry, and is further supported by the experimental data of Fig. 5.3 which demonstrate the expected relation between shrinkage and capillary pressure.

At some later stage, however, this pressure reaches a maximum and drops suddenly and rapidly. This maximum is sometimes referred to as breakthrough pressure and is attributed to the disruption in the continuity of the water system in the capillaries.

5.2.1. Factors Affecting Plastic Shrinkage

It was pointed out in the preceding section that the mechanism of plastic shrinkage is attributable to the tensile stresses in the capillary water which become operative when menisci are formed in the water in the capillaries on drying. It can be shown that this maximum tension occurs immediately below the surface and is equal to $2T/r$, where T is the surface tension of the water and r is the radius of curvature of the meniscus. The tension in the water increases with the decrease in the radius of curvature of the meniscus, whereas the latter decreases with the decrease in ambient relative humidity.† Accordingly, plastic shrinkage is expected to increase with the intensity of the drying conditions. It will be shown later (see section 5.2.1.1), that this is, indeed, the case.

It may be realised that the decrease in the radius of curvature, and the associated increase in the tension in the capillary water, may proceed only up to a certain point because the radius of curvature cannot be smaller than that of the capillary. Hence, on further drying the capillary is emptied and the tension is relieved explaining, in turn, the experimental data of Fig. 5.3. Accordingly, a maximum tension is reached (i.e. a breakthrough pressure) when the radius of the meniscus equals that of the capillary. It was suggested that this maximum capillary tension, P_c, is given by the following expression [5.3]:‡

$$P_c = kTSC/W \qquad (5.1)$$

where T is the surface tension of the water, S is the specific surface area of the cement, C is the cement content, W is the water content, and k is the ratio of the density of water to that of the cement. Accordingly, it is to be expected that the capillary pressure, and its associated plastic shrinkage, will increase with an increase in the cement content and its specific area, and decrease with an increase in the water content.

5.2.1.1. Environmental Factors

Environmental factors which affect drying include relative humidity, temperature and wind velocity. The effect of these factors is, of course, well known, and is clearly demonstrated in Fig. 5.4. In this respect it may be noted that, by far, the effect of the relative humidity is the most dominant (part A). The effect of the wind velocity (part B) is somewhat greater than that of temperature (part C) but is still much smaller than that of the relative humidity. In any case, in view of the suggested mechanism of plastic shrinkage, the latter is expected to increase with an increase in temperature and wind velocity and a decrease in relative humidity, through the effect of these

†The relationship between the radius of curvature, r, of the meniscus, and the corresponding vapour pressure, p, is given by Kelvin's equation $\ln(p/p_0) = 2T/R\theta\rho r$ where p_0 is the saturation vapour pressure over a plane surface (i.e. p/p_0 is the relative humidity), T is the surface tension of the water, R is the gas constant, θ is the temperature in K and ρ is the density of the water.

‡The expression $P_c = 0{\cdot}26TS\rho$, in which T is the surface tension of the water, S is the specific surface area of solid particles and ρ is their density, was also suggested [5.4].

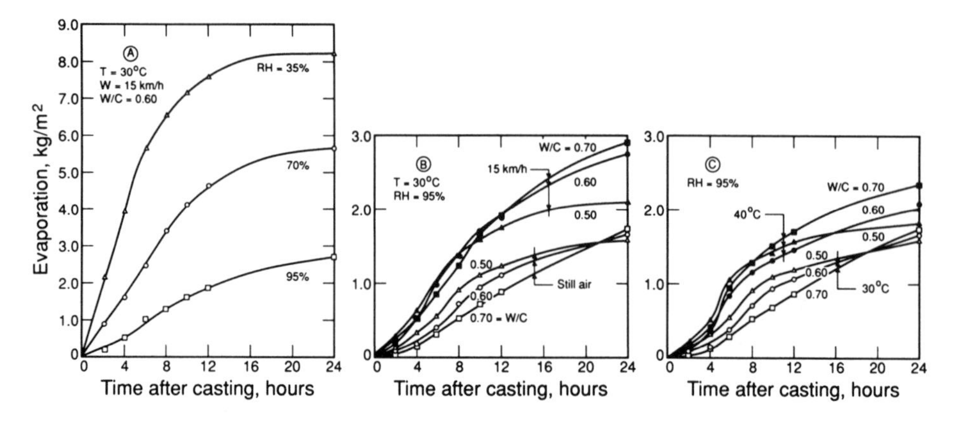

Fig. 5.4. Effect of (A) relative humidity, (B) wind velocity, and (C) ambient temperature on drying of fresh concrete. (Adapted from Ref. 5.5.)

environmental factors on the intensity of the drying process. In practice, however, this is not always the case, and plastic shrinkage is not necessarily the same for the same amount of water lost on drying (Fig. 5.5). This specific aspect is further dealt with in the following discussion.

Experimental data on the relation between plastic shrinkage and the

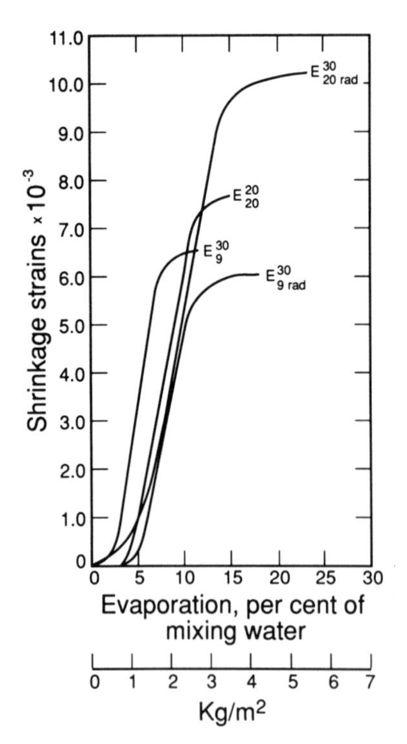

Fig. 5.5. Effect of evaporation on plastic shrinkage of cement mortars (plastic consistency, 550 kg/m^3 ordinary Portland cement (OPC)) subjected to different exposure conditions. Upper numbers refer to air temperature in centigrade, and lower numbers to wind velocity in km/h. 'rad' denotes exposure to IR irradiation. (Adapted from Ref. 5.6.)

intensity of drying of cement mortars, brought about by exposure to different environmental conditions, are presented in Fig. 5.5, where drying is measured by the amount of water loss. It may be noted, as can be expected from the preceding discussion, that, indeed, shrinkage increases with the increase in the amount of water lost, and this relation is essentially the same for all of the exposure conditions considered. On the other hand, ultimate shrinkage (i.e. total shrinkage which occurs until the concrete is set) differs considerably for the different exposure conditions. It can be seen, for example, that an increase in wind velocity from 9 to 20 km/h increased ultimate shrinkage from 6 to 9·7 mm/m (mixes E^{30}_{9rad} and E^{30}_{20rad} both exposed to IR irradiation at 30°C), whereas the amount of water lost remained virtually the same, i.e. some 20% of the mixing water. This difference is attributable to the simultaneous effect of the environmental factors on the stiffening rate and the setting time of the concrete. Ultimate shrinkage depends not only on the intensity of the drying, but also on the stiffness of the mix and the length of time it takes the mix to set, i.e. the stiffer the mix, and the shorter the setting time, the lower the expected shrinkage under otherwise the same conditions. The exposure conditions of mixes, E^{30}_{20rad} and E^{30}_{9rad}, differed only with respect to wind velocity. Consequently, the drying rate of mix E^{30}_{20rad} was greater than of mix E^{30}_{9rad} but the setting time of both mixes was essentially the same. That is, a greater part of the drying of mix E^{30}_{20rad} took place at an earlier age, when the mix was less rigid than mix E^{30}_{9rad}. Hence, the higher ultimate shrinkage exhibited by the former mix. In other words, ultimate shrinkage is determined quantitatively by the net effect of the environmental factors on both the rate of drying and rate of setting.

In view of the preceding discussion, it may be expected that the use of set-retarding admixtures will increase plastic shrinkage and, indeed, this is confirmed by the data of Fig. 5.6, which compare the shrinkage of retarded and non-retarded cement mortars which were otherwise the same. An increased plastic shrinkage is associated with an increased risk of plastic cracking. Hence, the use of retarders should preferably be avoided under environmental conditions, such as hot, dry weather conditions, which favour high plastic shrinkage. This conclusion is of practical importance because the use of retarders is sometimes recommended under hot, dry conditions in order to counteract the accelerated effect of such conditions on slump loss in fresh concrete (section 4.3.2).

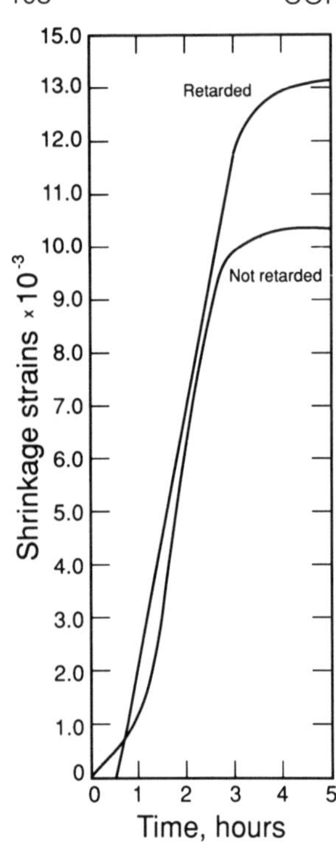

Fig. 5.6. Plastic shrinkage of retarded and unretarded cement mortars of plastic consistency and OPC content of 550 kg/m^3. Air temperature of 30°C, wind velocity of 20 km/h and IR irradiation. (Adapted from Ref. 5.6.)

5.2.1.2. Cement and Mineral Admixtures

It was pointed out earlier (section 5.2.1) that in accordance with eqn (5.1) for the capillary pressure, the latter is expected to increase with an increase in the cement content and its fineness (i.e. specific surface area). In fact, such a trend is to be expected because the greater the cement content, the greater the number of contact points at which the menisci are formed and the capillary tension becomes operative. Similarly, the smaller the size of the cement grains, the smaller the radii of the menisci which are formed at the contact points. Consequently, under otherwise the same conditions, a greater capillary tension is expected with an increase in the cement content and its fineness, and, similarly, the associated plastic shrinkage is expected to increase as well. Strictly speaking, in this respect all the granular ingredients of the concrete mix should be considered. The size of the aggregate particles, however, is many times greater than that of the cement grains, and their effect on the capillary tension is of no significance at all. Hence, in this respect, only the cement content matters. On the other hand, the cement content should be extended to include mineral admixtures which have a specific surface area of the same order of that of the cement (e.g. fly-ash) or greater (e.g. microsilica). The effect of the cement content on plastic shrinkage is clearly demonstrated in Fig. 5.7.

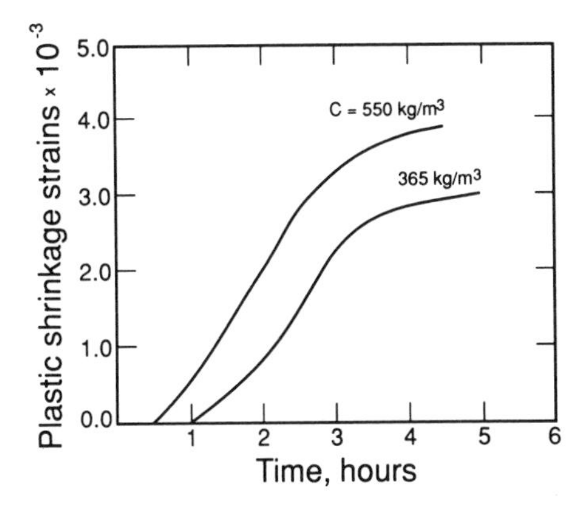

Fig. 5.7. Effect of the cement content on plastic shrinkage of cement mortars of semi-plastic consistency. Air temperature 30°C, RH 45%, wind velocity 20 km/h. (Adapted from Ref. 5.7.)

The plastic shrinkage of fly-ash concrete is compared in Fig. 5.8 to that of a similar concrete made without fly-ash. In the mixes tested 20% of the cement was replaced by fly-ash. However, in order to facilitate comparison at the same strength level, each 1 kg cement was replaced by 1·7 kg fly-ash. Consequently, the cement + fly-ash content in the fly-ash concrete was 14% greater than the cement content in the reference concrete. Due to the greater combined

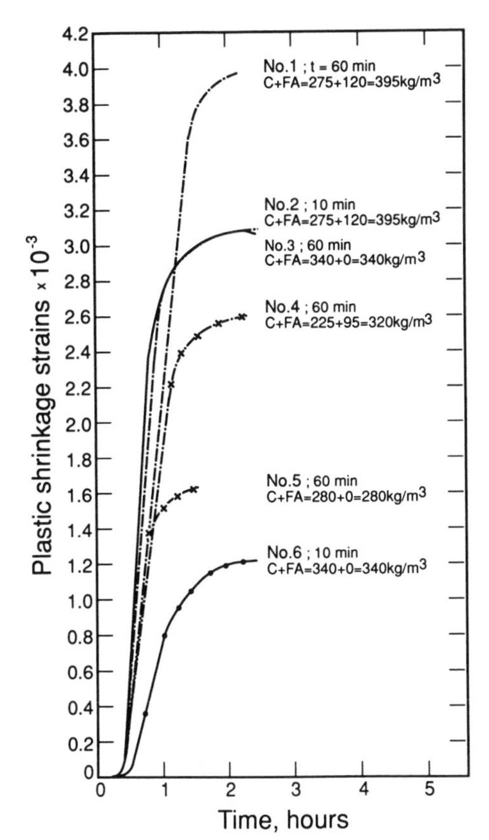

Fig. 5.8. Effect of the fly-ash addition, mixing time and cement content on plastic shrinkage of concrete. (Adapted from Ref. 5.8.)

cement + fly-ash content, the fly-ash concrete should exhibit a greater plastic shrinkage than the reference concrete. This is clearly evident from Fig. 5.8 when the shrinkage curves are compared for the same mixing time and original cement content, i.e. curves 4 and 5 (60 min mixing time, 280 kg/m^3 cement), 1 and 3 (60 min mixing time, 340 kg/m^3 cement), and 2 and 6 (10 min mixing time, 340 kg/m^3 cement). In fact, the effect of fly-ash was quite significant, increasing, in the case of 10 min mixing, plastic shrinkage by approximately a factor of three (compare curves 2 and 6). It should be realised that this effect of the fly-ash on plastic shrinkage is also partly attributable to its delaying effect on the setting of the fresh concrete. Hence, the length of time in which plastic shrinkage takes place is longer in fly-ash concrete than in its ordinary counterpart and, therefore, a greater shrinkage is expected in the former than in the latter concrete.

It is also evident from Fig. 5.8 that plastic shrinkage increases significantly with an increase in mixing time from 10 to 60 min (compare curves 1 and 2, and 3 and 6). This increased shrinkage is attributable to the grinding effect of the mixing operation which, on prolonged mixing, increases the fines content in the concrete mix.

Finally, the data of Fig. 5.8 also fully support the previous conclusion that a greater cement content involves a greater shrinkage (compare curves 3 and 5).

It was pointed out earlier (see section 3.1.2.2.2) that microsilica has an average grain size of 0·1 μm, as compared with an average size of 10 μm for Portland cement. Hence, it is to be expected that incorporating microsilica in the concrete mix will increase significantly plastic shrinkage. Data directly relating to this expected effect are not available, but it was observed that the addition of microsilica having a specific surface area of 23 900 m^2/kg significantly increased plastic cracking [5.9].

5.2.1.3. Water Content

In accordance with eqn (5.1), capillary pressure is expected to decrease with an increase in the water content in the concrete mix and, accordingly, a lower shrinkage is to be expected in a wet mix than in its dry counterpart. In practice, however, the opposite behaviour is observed, namely, that plastic shrinkage is greater in wet than in dry mixes (Fig. 5.9). Moreover, such behaviour is indirectly supported by the observation that plastic cracking did not occur under severe evaporation conditions in semi-plastic mortars, while plastic and wet mortars, of the same dry mix proportions, cracked severely [5.10]. Again, this apparent contradiction between the expected and the observed shrinkage,

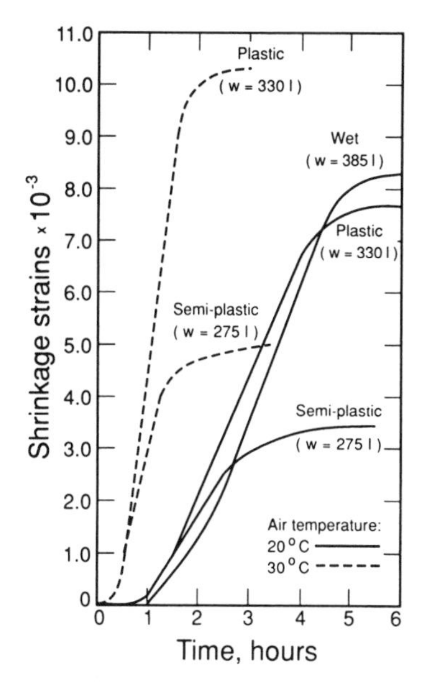

Fig. 5.9. Effect of water content (w) on plastic shrinkage of cement mortars with OPC content of 550 kg/m^3 at different exposure conditions. (Adapted from Refs 5.6 and 5.10.)

may be attributed to the effect of the water content on the stiffness of the mortar, and thereby on its plastic shrinkage. A lower water content results in a stiffer mix which, in turn, resists shrinkage to a greater extent than a wetter mix with a higher water content. Apparently, this effect of the water content is greater than its expected theoretical effect and, consequently, plastic shrinkage of wet mortars is greater than that of dry ones.

5.2.1.4. Chemical Admixtures

Chemical admixtures (see section 4.3.2) affect plastic shrinkage through their effect on water content and setting time. Accordingly, water-reducing admixtures are expected to reduce shrinkage due to the reduced water demand involved in their use, whereas the use of set-retarding admixtures is expected to increase shrinkage due to their delaying effect on setting of concrete. This expected effect is confirmed by the data of Fig. 5.6, and is discussed in some detail in section 5.2.1.1.

5.2.1.5. Fibre Reinforcement

Fibres are sometimes incorporated in concrete mainly to increase its toughness, and thereby improve its performance under impact and dynamic loading. In some cases, but not always, concrete tensile strength is improved as well. Different types of fibres are available but, at present, steel, polypropylene and

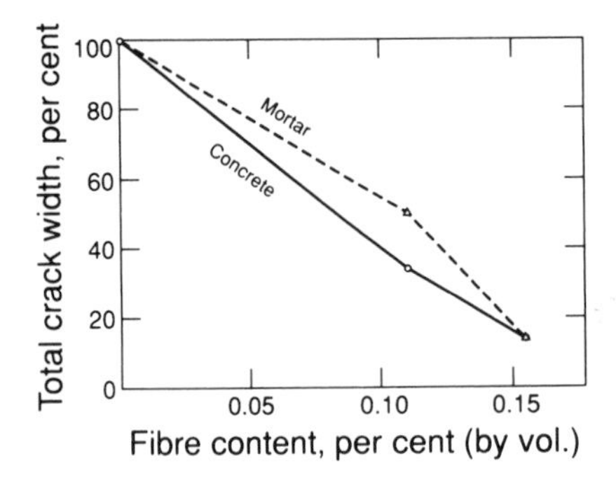

Fig. 5.10. Effect of volume concentration of polypropylene fibres on total width of cracks induced by restrained plastic shrinkage. (Adapted from Ref. 5.14.)

glass-fibres are mostly used, the former two mainly on the building site, and the latter mainly in the production of glass-fibre-reinforced concrete products, commonly known as GRC products. A detailed discussion of fibre-reinforced concrete can be found, for example, in Ref. 5.11.

Steel fibres, due to their restraining effect, were shown to reduce plastic shrinkage [5.12], and this effect increased with the increase in the product of their volume concentration, v_f, and aspect ratio, l/d.† As well, fibres which due to their configuration, are of better bond properties (e.g. crimpled fibres), further reduce shrinkage suggesting, once again, that the effect of the fibres is due to a restraining mechanism. Moreover, the use of polypropylene fibres [5.13] and glass-fibres (Bentur, A., pers. comm.) has been shown to eliminate plastic cracking or to reduce it considerably (Fig. 5.10). Hence, the incorporation of fibres in the concrete mix may be considered an efficient means to control plastic cracking. Indeed, polypropylene fibres are increasingly used to control plastic shrinkage cracking, at fibre addition rates of 0·1% by volume.

5.2.2. Plastic Shrinkage Cracking

It was already explained that plastic cracking occurs when the tensile stress in the not yet hardened concrete, brought about by the restrained shrinkage, exceeds its tensile strength. The occurrence of cracking depends, sometimes, on contradictory factors, and cannot be related directly either to the intensity of drying (i.e. water loss) or to the amount of shrinkage. Nevertheless, it may be generally stated that the likelihood of plastic cracking increases with the intensity of the drying, on the one hand, and decreases with the increase in the rate of stiffening and strength development of the fresh concrete, on the other.

†l is the fibre's length and d is its diameter or, in the case of non-circular cross-section, the equivalent diameter. The latter equals $\sqrt{4A/\pi}$ where A is the cross-sectional area of the fibre.

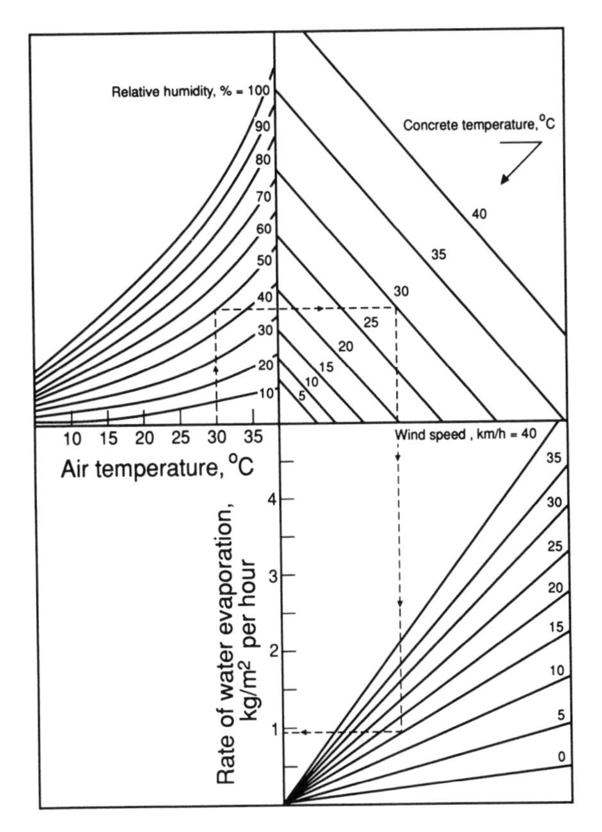

Fig. 5.11. Effect of climatic factors on the rate of evaporation from fresh concrete. (Adapted from Ref. 5.15.)

It follows that all factors which affect drying, plastic shrinkage and setting, will similarly affect the likelihood of cracking.

In practice, it is very difficult, if not impossible, to consider all the factors involved in order to determine the possibility of plastic cracking to occur under a given situation. It has been suggested, however, that when the rate of evaporation from the fresh concrete approaches $1{\cdot}0\,kg/m^2$ per hour, 'precautions against plastic shrinkage are necessary', and in order to estimate this rate under the expected climatic conditions, a suitable chart has been provided (Fig. 5.11). It is recommended, however, that in hot climates, and particularly in hot, dry climates, plastic cracking should be always considered as a distinct possibility, and suitable means employed in order to prevent such cracking. Noting that plastic shrinkage is conditional on drying, cracking can be prevented simply by protecting the concrete as early as possible, and always before its surface dries out. Such protection can be achieved by covering the

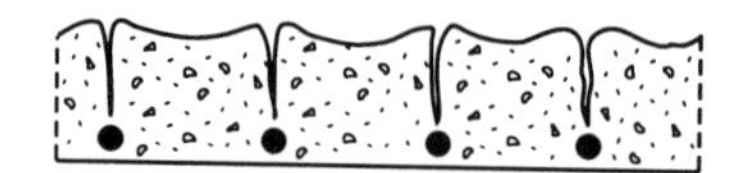

(A) Obstruction of settlement due to the presence of reinforcing bars

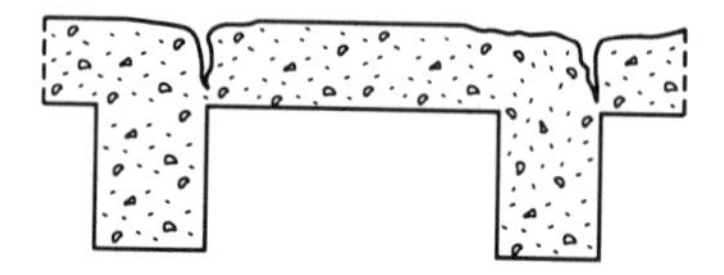

(B) Differential settlement due to geometry of cross-section

Fig. 5.12. Schematic description of plastic settlement cracking.

concrete with, say, polyethelene sheeting or, under moderate conditions, spraying its surface with a suitable sealing compound. Experience has shown that these means, when adequately applied, are usually successful in preventing plastic cracking. As mentioned earlier, the incorporation of fibres in the concrete mix may be also useful.

5.3. PLASTIC SETTLEMENT AND CRACKING

When concrete is placed, there exists a tendency for the water in the mix to rise to the surface, and for the solids to settle, i.e. bleeding occurs. Excessive bleeding is characteristic of wet mixes deficient in fines. On the other hand, increased fineness of the cement, and replacing part of the sand with a fine filler, both reduce bleeding. Accelerating admixtures reduce the time during which the concrete remains plastic and can settle, and thereby reduce bleeding. Air entrainment is also very effective in reducing bleeding and its associated settlement.

Plastic settlement cracks occur in concrete which exhibits a relatively high bleeding and has its settlement obstructed by, for example, the presence of reinforcing bars or the geometry of the cross-section of the element involved (Fig. 5.12). Such obstructions cause differential settlement which, in turn, may cause cracking if it occurs when the concrete is brittle and weak and cannot accommodate such settlement. It follows that, unlike plastic shrinkage cracks, settlement cracks are orientated and follow reinforcing bars and other obstructions, as the case may be.

Plastic settlement cracks, if they occur, can be eliminated by revibration of the concrete provided, of course, the concrete is still plastic enough to allow such a revibration. Using a trowel or a float to close the cracks may be adequate when thin sections are involved. In thick sections, however, the cracks may re-open on drying of the hardened concrete, because trowelling and floating are, essentially, surface treatments which do not affect the deeper parts of the cracks.

5.4. SUMMARY AND CONCLUDING REMARKS

Plastic cracking occurs when the fresh concrete is exposed to a high rate of drying (evaporation) at the stage when it is brittle and not strong enough to resist the tensile stresses induced by the restrained plastic shrinkage. Hence, all factors which accelerate drying, i.e. higher ambient temperatures, greater wind velocity, and lower relative humidity, increase the likelihood of plastic cracking. Accordingly, the occurrence of plastic cracking must be considered as a distinct possibility under hot, and particularly under hot, dry weather conditions. The likelihood of cracking is further increased with the use of cement-rich and wet mixes, and with the use of mineral and set retarding admixtures. On the other hand, fibre-reinforcement virtually eliminates plastic cracking. Plastic cracking can be effectively controlled by protecting the fresh concrete from drying as early as possible, but always before its surface dries out. Covering the concrete with polyethelene sheeting or spraying its surface with a suitable sealing compound, are both adequate means to protect the concrete against plastic cracking.

Plastic settlement cracking occurs when the settlement of concrete, which is characterised by high bleeding, is obstructed. Wet mixes, retarded mixes and those deficient in fines are more sensitive to settlement cracking. Air entrainment, however, reduces bleeding considerably, and thereby settlement cracking as well. In general, the adequate and early protection from drying of a well-designed concrete is usually enough to prevent the occurrence of plastic settlement cracks. Further prevention can be achieved by the use of air-entraining admixtures.

REFERENCES

5.1. Soroka, I. & Jaegermann, C., Deterioration and durability of concrete in hot climates. In *Proc. RILEM Seminar on Durability of Concrete Structures Under Normal Outdoor Exposure*. Universitat Hannover, Hannover, 1984, pp. 52–64.

5.2. Wittmann, F.H., On the action of capillary pressure in fresh concrete. *Cement Concrete Res.*, **6**(1) (1976), 49–56.

5.3. Powers, T.C., Physical properties of cement paste. In *Proc. Symp. Chem. of Cement* (Vol. II), Washington, 1960, pp. 577–613.

5.4. Pihlajavaara, S.E., A review of the main results of a research on the aging phenomena of concrete: Effect of moisture conditions on strength, shrinkage and creep of mature concrete. *Cement Concrete Res.*, **4**(5) (1974), 761–71.

5.5. Shalon, R. & Berhane, Z., Shrinkage and creep of mortar and concrete as affected by hot humid environment. In *Proc. RILEM 2nd Int. Symp. on Concrete and Reinforced Concrete in Hot Countries*, Haifa, 1971, Vol. II, Building Research Station–Technion, Israel Institute of Technology, Haifa, pp. 309–32.

5.6. Ravina, D. & Shalon, R., Shrinkage of fresh mortars cast under and exposed to hot dry climate conditions. In *Proc. RILEM/CEMBUREAU Colloq. on Shrinkage of Hydraulic Concretes*. Madrid, 1961, Vol. II, Edigrafis, Madrid.

5.7. Ravina, D., The mechanism of plastic cracking of concrete. PhD thesis, Faculty of Civil Engineering, Technion—Israel Institute of Technology, Haifa, Israel, August 1966 (in Hebrew with an English summary).

5.8. Ravina, D. & Jaegermann, C., Effect of partial replacement of the cement by fly ash on plastic cracking tendency of concrete in hot weather. Research Report 017-401, Building Research Station, Technion—Israel Institute of Technology, Haifa, Israel, Oct. 1986 (in Hebrew).

5.9. Cohen, M.D., Olek, J. & Dolch, W.L., Mechanism of plastic shrinkage cracking in Portland cement and Portland cement–silica fume paste and mortar. *Cement Concrete Res.*, **20**(1) (1990), 103–19.

5.10. Ravina, D. & Shalon, R., Plastic shrinkage cracking. *J. ACI*, **65**(4) (1968), 282–92.

5.11. Bentur, A. & Mindess, S., *Fibre Reinforced Cementitious Composites*. Elsevier Applied Science, London, UK, 1990.

5.12. Mangat, P.S. & Azari, M.M., Plastic shrinkage of steel fibre reinforced concrete. *Mater. Struct.*, **23**(135) (1990), 186–95.

5.13. Al-Tayyib, A.J., Al-Zahrani, M.M., Rasheeduzzafar & Al-Sulaimani, G.J., Effect of polypropylene fiber reinforcement on the properties of fresh and hardened concrete in the Arabian Gulf environment. *Cement Concrete Res.*, **18**(4) (1988), 561–70.

5.14. Dahl, A.P., Influence of fibre reinforcement on plastic shrinkage cracking. In

Brittle Matrix Composites (Proc. European Mechanics Colloquium 204), ed. A.M. Brandt & I.H. Marshall. Elsevier Applied Science, London, UK, 1986, pp. 435–41.

5.15. ACI Committee 305, Hot weather concreting (ACI 305, R-89). In *ACI Manual of Concrete Practice* (Part 2). ACI, Detroit, MI, USA, 1990.

Chapter 6
Concrete Strength

6.1. INTRODUCTION

Concrete may be regarded as a composite material in which the hardened cement paste constitutes the continuous phase, and the aggregate particles, which are embedded in the paste, are the discrete phase. Accordingly, it may be surmised that concrete strength will be determined by the strengths of the hardened cement paste and the aggregates, the strength of the paste–aggregate bond and the aggregate concentration in the paste, i.e. the aggregate content in the concrete. This is, indeed, the case but, as it will be seen later, the effect of some of these strength-determining factors is comparatively small and it is ignored, therefore, in everyday practice.

6.2. STRENGTH OF HARDENED CEMENT PASTE

It was shown earlier (Section 2.4) that the hardened paste is characterised by a highly porous structure and, in fact, porosity constitutes the most dominant factor with respect to its strength. As in other porous solids, the relation between the strength of the paste, S, and its porosity, p, may be generally expressed by

$$S = S_0 \exp(-bp) \tag{6.1}$$

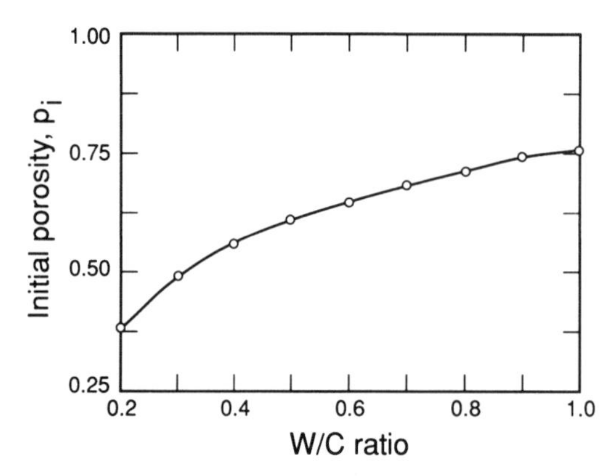

Fig. 6.1. The relation between initial porosity and W/C ratio in a cement paste in accordance with eqn (6.2) ($V_c = 0{\cdot}32\,cm^3/g$).

where S_0 is the strength of the paste at zero porosity (i.e. $p = 0$) and b is a constant which depends on the type of the cement involved, age of the paste, etc. It must be realised that other expressions have been suggested to describe the strength–porosity relationship in the hardened cement paste. Nevertheless, and regardless of the exact nature of the relationship in question, strengthwise, porosity remains the most dominant factor. Hence, all factors which determine the porosity of the paste determine its strength as well. In this respect the water to cement (W/C) ratio and the degree of hydration are the main factors involved.

6.2.1. Effect of W/C Ratio on Initial Porosity

The W/C ratio determines the initial distance between the unhydrated cement grains in the water–cement mix, i.e. it determines the relative water content in the mix. Accordingly, the latter is sometimes referred to as the 'initial' porosity of the paste. It can be shown that the initial porosity, p_i, of the paste is given by

$$p_i = \omega/(V_c + \omega) \tag{6.2}$$

where V_c is the specific volume of the cement and ω is the W/C ratio. Equation (6.2) is plotted in Fig. 6.1 from which it is clearly evident that the initial porosity increases with the increase in the W/C ratio.

6.2.2. Combined Effect of W/C Ratio and Degree of Hydration on Porosity

It was pointed out earlier (Section 2.4) that the volume of the hydration products is some 2·2 times greater than the volume of the unhydrated cement.

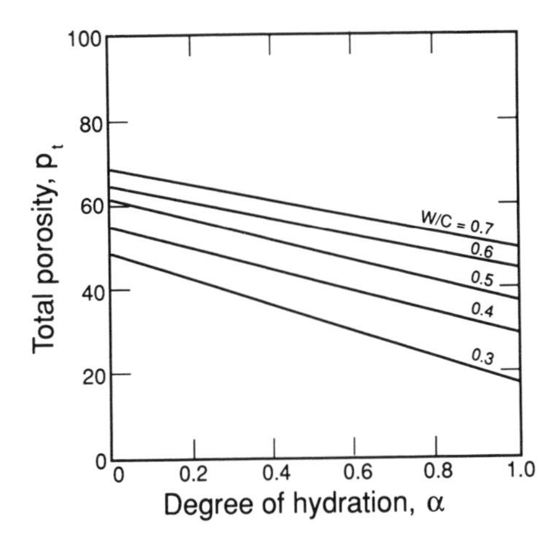

Fig. 6.2. The effect of W/C ratio and degree of hydration on total porosity of cement paste.

Consequently, the spacing between the cement grains, and the porosity of the paste, both decrease as the hydration proceeds. That is, at a given stage, the porosity of the paste, and its associated strength, are determined by both the W/C ratio and the degree of hydration. It can be shown that the combined effect of the W/C ratio and the degree of hydration on porosity is given by eqn (6.3), assuming the volume of the solids is increased by the factor of 2·2 and the specific volume of the cement is 0·32 cm^3/g [6.1]:

$$p_t = 1 - \frac{0{\cdot}32 + 0{\cdot}187\alpha}{0{\cdot}32 + \omega} \qquad (6.3)$$

in which p_t is the total porosity of the paste (i.e. the combined volume of gel and capillary pores), ω the W/C ratio and α is the degree of hydration. This expression is presented in Fig. 6.2, clearly indicating the expected decrease in porosity with the increase in the degree of hydration and the decrease in the W/C ratio.

6.2.3. Effect of W/C Ratio on Strength

It follows from the preceding discussion that, for the same degree of hydration (i.e. for the same age and curing regime), the porosity of the paste is determined by the W/C ratio alone. If, indeed, porosity determines strength, it may be further stipulated that under the same conditions, strength, as well, will be determined by the W/C ratio alone. The experimental data presented in Fig. 6.3 fully corroborate the latter stipulation and, indeed, it is generally recognised and accepted.

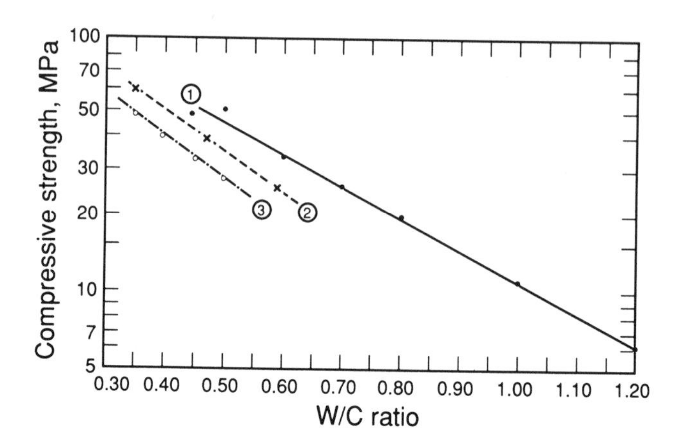

Fig. 6.3. Relation between the compressive strength of cement pastes and W/C ratio. (Adapted from the data of (1) Soroka, I. & Sereda, J.P., unpublished data, 1967, (2) Ref. 6.2, and (3) Ref. 6.3.)

6.3. STRENGTH OF PASTE–AGGREGATE BOND

The bond between the cement paste and the embedded aggregate particles is due to mechanical and physical effects and, apparently, but to a lesser extent, to chemical reactions which may take place between the cement and the aggregate. In practice, however, the main factors involved are the W/C ratio and surface characteristics of the aggregate particles.

6.3.1. Effect of W/C Ratio

The effect of W/C ratio on the strength of the paste–aggregate bond is similar to its effect on the strength of the paste, i.e. a decrease in the W/C ratio simultaneously increases the strength of the paste, as well as the strength of the paste–aggregate bond [6.4, 6.5]. Concrete strength is mainly determined by the strength of the paste and the strength of its bond to the aggregate explaining, in turn, why the W/C ratio is the most important factor with respect to concrete strength.

6.3.2. Effect of Surface Characteristics

It is to be expected that a rougher aggregate surface would improve the bond and, consequently, would result in a concrete of a higher strength. Indeed,

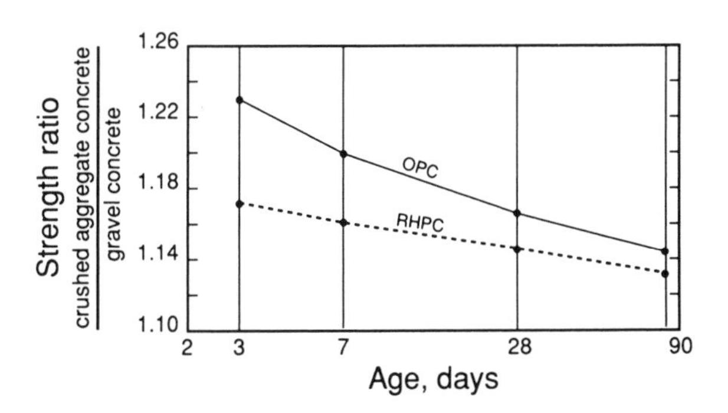

Fig. 6.4. Approximate strength ratio of crushed aggregate concrete to gravel aggregate concrete. W/C ratio of 0·5. (Adapted from Ref. 6.7.)

experience, as well as experimental data [6.6], have shown that the strength of concrete made with crushed aggregate is stronger than otherwise the same concrete made with gravel. Generally, this effect is greater in the lower than in the higher W/C ratio range, and may disappear completely in low strength concretes. This effect on the compressive strength of concrete is indicated, for example, by the data presented in Fig. 6.4. Similarly, reductions have been observed in flexural strength [6.36] and, apparently, the latter strength is even more sensitive to surface roughness than compressive strength.

6.3.3. Effect of Chemical Composition

Generally speaking, common concrete aggregates are considered to be inert in the water–cement system. However, there exist some experimental data which indicate that, in some aggregates, chemical reactions take place at the paste–aggregate interface [6.8–6.11], and a distinctive layer, different in composition from both the paste and the aggregate, is formed at the interface. It is not exactly clear to what extent the formation of this layer contributes to the paste–aggregate bond, and what is the chemical and mineralogical composition of the aggregate which favours its formation. In daily practice, however, neither the formation of such a layer, nor the composition of the aggregates, are considered with respect to concrete strength.

6.3.4. Effect of Temperature

There exist some limited experimental data, presented in Fig. 6.5, which indicate that the strength of paste–aggregate bond is independent of curing

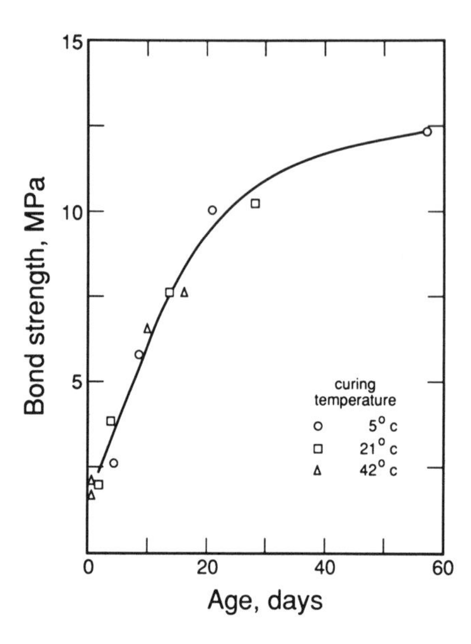

Fig. 6.5. The development of paste–aggregate bond strength and its independence of curing temperature. (Adapted from Ref. 6.4.)

temperature. This is a somewhat unexpected observation because temperature affects the rate of hydration (see section 2.5.1), whereas the latter, through its effect on the porosity of the paste, affects concrete strength. Accordingly, it is to be expected that bond strength would increase with temperature. Hence, the observed independence of bond strength on curing temperature implies that some adverse effect is involved simultaneously which counteracts the expected improvement in strength with temperature. Such an adverse effect may be attributed to cracking at the paste–aggregate interface, which is brought about by differential thermal volume changes of the two phases involved, i.e. the coefficient of thermal expansion of cement paste, depending on its moisture content, varies from 11 to 20 $\times$ 10^{-6} per °C and that of normal aggregates is usually lower and varies, in most cases, from 5 to 13 $\times$ 10^{-6} per °C [6.12].

6.4. EFFECT OF AGGREGATE PROPERTIES AND CONCENTRATION ON CONCRETE STRENGTH

The discussion in the preceding section was limited to aggregate properties which affect concrete strength through their effect on paste–aggregate strength. Some other properties, which indirectly affect concrete strength through their effect on water demand, were also discussed earlier (see section 4.2.1). It may be expected, however, that some additional properties of the

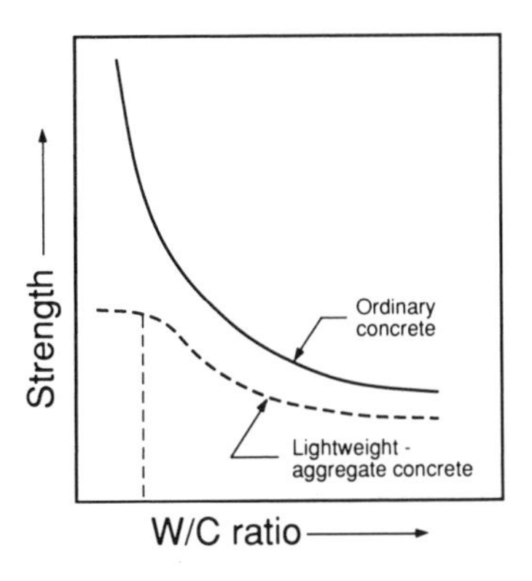

Fig. 6.6. The difference in strength of lightweight and normal-weight aggregate concretes of the same W/C ratio.

aggregate, such as strength, modulus of elasticity, etc., as well as its concentration in the hardened concrete, will also affect concrete strength. In this respect, it must be realised that it is rather difficult to differentiate quantitatively between the effects of aggregate properties on concrete strength. This difficulty stems from the fact that aggregate properties, such as, for example, strength and modulus of elasticity, change simultaneously. Hence, it is rather difficult, if not impossible, to separate experimentally their effect on concrete strength.

6.4.1. Effect of Aggregate Strength

Since concrete is a composite material, it is to be expected that its strength will be affected by that of the aggregate. Indeed, generally, for the same W/C ratio concretes made of lightweight aggregates are weaker than those made of normal-weight aggregates. This difference in strength, which is schematically described in Fig. 6.6, may be attributed to the lower strength of the lightweight aggregates. The lower strength of the latter aggregates also affects the mode of concrete failure, i.e. in lightweight aggregate concrete fracture extends throughout the aggregate (Fig. 6.7(A)), whereas in normal-weight concrete it occurs mostly at the paste–aggregate interface (Fig. 6.7(B)). This mode of failure explains the existence of a limiting strength in lightweight aggregate concrete, which is not increased by further reductions in the W/C ratio (Fig. 6.6). Apparently, at this strength level, the strength of the aggregate is lower than that of the paste. Consequently, concrete strength is mostly controlled by that of the aggregate and any further increase in the strength of the paste, brought about by the reduced W/C ratio, does not result, therefore, in a significant effect on concrete strength. In normal-weight concrete, in which the aggregate is much stronger than the paste, this is not the case, and concrete strength increases with the decrease in the W/C ratio as long as it can

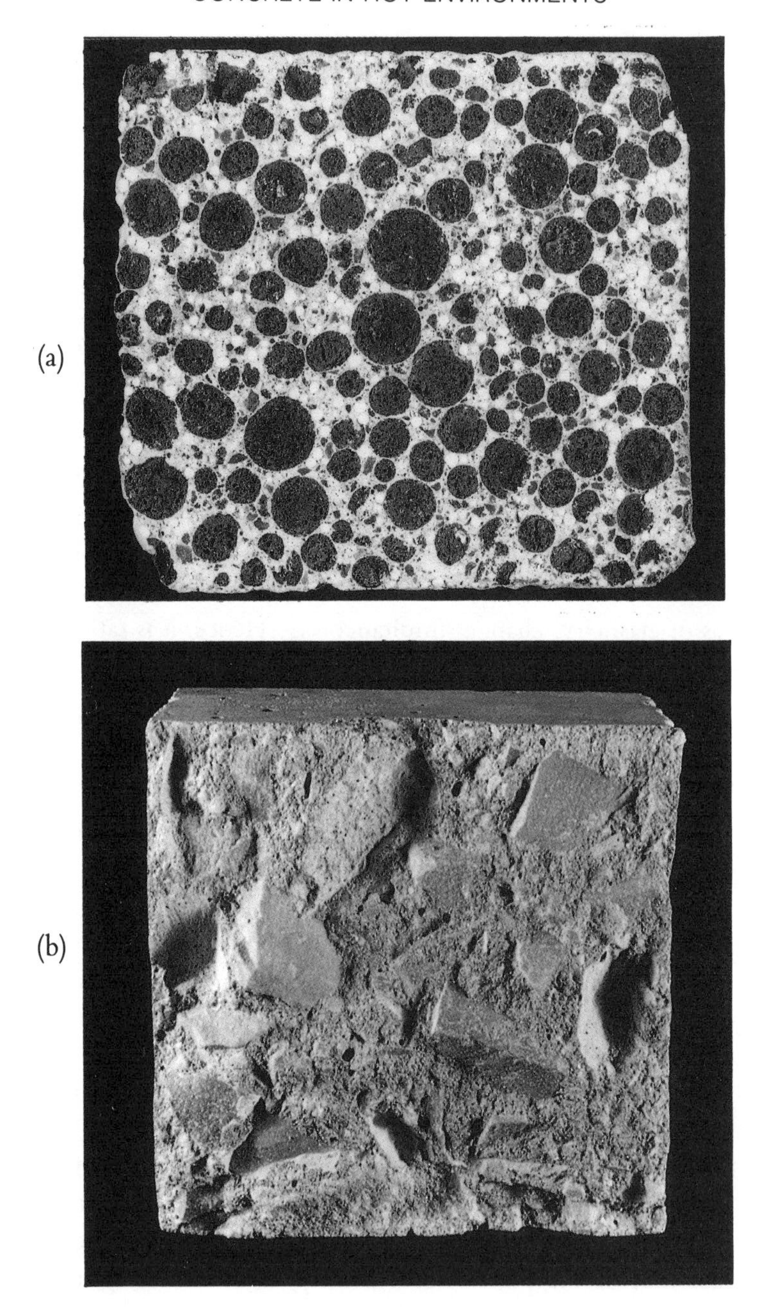

Fig. 6.7. Mode of failure in (A) lightweight aggregate (expanded clay) concrete, and (B) normal-weight aggregate (crushed dolomite) concrete.

be adequately compacted. That is, in normal concrete strength is determined mainly by that of the paste and, for practical purposes, it is generally assumed that in such concrete aggregate strength hardly affects concrete strength. On the other hand, when high strength concrete is involved (i.e. compressive strength in the order of 100 MPa and higher), the strength of the aggregate again becomes very important as does its bond to the cement paste.

6.4.2. Effect of Aggregate Modulus of Elasticity

The modulus of elasticity of the aggregate constitutes one of the factors which determines concrete strength, and in general concrete strength increases with an increase in modulus of elasticity of the aggregate [6.6, 6.13]. The latter relation between concrete strength and modulus of elasticity of the aggregate may be explained from the effect of aggregate rigidity on stress distribution in concrete under load. Assuming equal strains, the part of the load which is taken by the aggregate increases with its rigidity (i.e. its modulus of elasticity), and, consequently, the part taken by the paste decreases. In ordinary concrete, in which the aggregate is significantly stronger than the paste, strength is determined mainly by the strength of the paste. Hence, the decrease in the load which is taken by the paste delays fracture and thereby increases concrete strength. In this respect, it should be noted that a higher modulus of elasticity characterises a stronger aggregate. It follows that the strength differences, indicated in Fig. 6.6, are actually due to differences in both aggregate strength and modulus of elasticity.

6.4.3. Effect of Particle Size

The presence of aggregate particles in the cement paste induces stress concentrations at, and close to, the paste–aggregate interface. The greater the particle size, the higher the stress concentration, thereby causing an earlier failure. Hence, it is to be expected that concrete strength will decrease with an increase in maximum particle size. This conclusion is supported by many findings [6.14–6.17] and is demonstrated, for example, by the experimental data presented in Fig. 6.8. It may be noted, however, that this effect of particle size is comparatively small. Hence, it is usually ignored in everyday practice where the maximum particle size varies within the narrow range of, say, 20–37·5 mm.

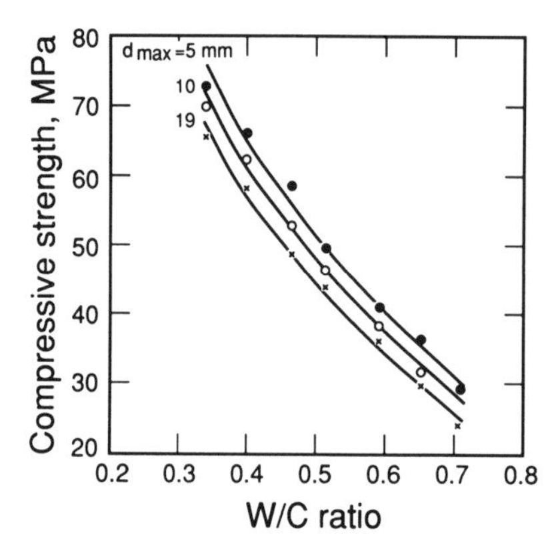

Fig. 6.8. Effect of aggregate particle size on concrete strength. (Adapted from Ref. 6.18.)

6.4.4. Effect of Aggregate Concentration

Although there exist data to the contrary [6.19], it is generally accepted that concrete strength increases with an increase in aggregate concentration. As has already been mentioned, the presence of the aggregate particles within the paste induces stress concentrations. The regions of stress concentration around neighbouring aggregates overlap, and this overlapping of regions increases as the aggregate concentration is increased. Consequently, the average stress concentration induced by the aggregate particles decreases as the concentration of the aggregate increases, and concrete strength is increased. This conclusion is apparent in Fig. 6.9 and is supported by other data [6.14, 6.21, 6.22]. In this respect, it should be noted that the preceding conclusion is valid provided that the reduced paste content remains high enough to allow complete consolidation of the concrete. Otherwise, the reduced paste content will result in a voids-containing concrete which, therefore, will be weaker. That is, an optimum aggregate concentration is to be expected with respect to concrete strength. In mortars this optimum was found to be 30 and 36% for compres-

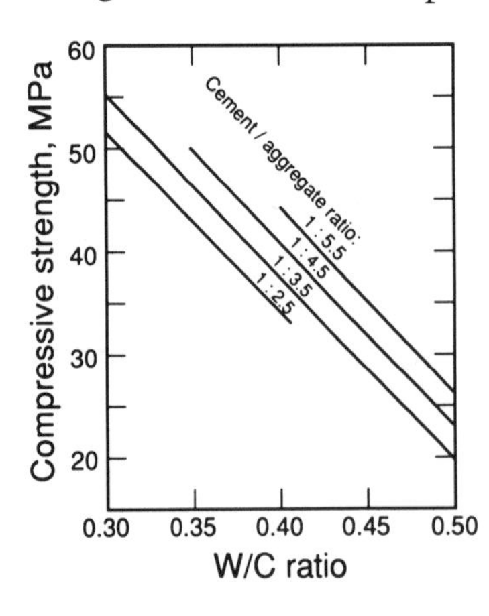

Fig. 6.9. Effect of aggregate concentration on strength of concrete. (Adapted from Ref. 6.20.)

sive and flexural strength, respectively [6.23, 6.24]. In any case, aggregate concentration in normal concrete varies within the narrow range of, say, 65–75%. Consequently, the variation in its effect on concrete strength is comparatively small and is usually ignored in daily practice.

6.4.5. Summary

Many aggregate properties affect concrete strength and, generally, concrete strength increases with an increase in aggregate modulus of elasticity, strength and concentration and decreases with its particle size. The effect of these properties is, however, comparatively small and is not usually considered in daily practice. More significant in this respect are the surface characteristics of the aggregate which affect the strength of the paste–aggregate bond, and thereby concrete strength, and particularly its flexural strength. Indeed, strengthwise, this effect of surface characteristics is recognised in mix design, and to this end a distinction is made between crushed and uncrushed (gravel) aggregate. As can be expected, and as it was pointed out earlier (Fig. 6.4), the use of smooth and round aggregate (gravel) results in a lower strength than the use of rough and angular (i.e. crushed) aggregate.

6.5. STRENGTH–W/C RATIO RELATIONSHIP

It was shown in the preceding discussion that porosity determines the strength of the cement paste, whereas, in turn, porosity is determined by the W/C ratio and the degree of hydration. That is, for the same degree of hydration, the difference in strength of pastes is determined solely by the W/C ratio. Concrete strength, in turn, is determined not only by the strength of the cement paste, but also by the strength of the paste–aggregate bond and by some properties of the aggregate. In this respect, however, and particularly when the compressive strength is considered, the strength of the paste is the main factor. The strength of the paste, as well as that of the paste–aggregate bond, are mainly determined by the W/C ratio. Hence, for the same degree of hydration (i.e. for the same type of cement, age and curing conditions), and the same normal-weight aggregate (in practice only of the same surface characteristics), concrete strength is determined by the W/C ratio alone and can be expressed

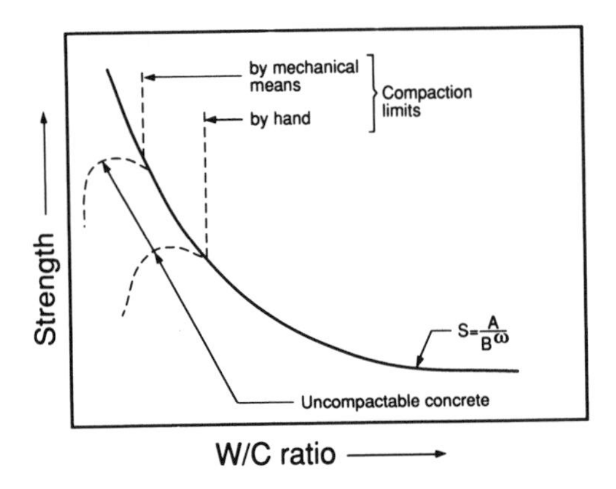

Fig. 10. Schematic description of the relation between W/C ratio and concrete strength.

by the expression

$$S = A/B^{\omega} \tag{6.4}$$

where S is concrete strength and ω is the W/C ratio; A and B are constants which depend on the remaining factors which affect strength such as, curing regime, type of cement and surface characteristics of the aggregate.

This relation between strength and W/C ratio is sometimes referred to as 'Abrams' law' [6.25]. Actually, this expression resulted from curve-fitting of experimental data and, strictly speaking, it is not a 'law'. At other times this expression is referred to as the 'W/C ratio law'.

The W/C ratio law is widely applied in mix design. Once the constants A and B are determined for the conditions in hand, the resulting curve can be used to estimate concrete strength from the W/C ratio or, alternatively, to select the W/C ratio required to produce a concrete of a desired strength. Indeed, in this context, the W/C ratio law plays an important role in concrete mix design.

The W/C ratio law (eqn (6.4)) is schematically described in Fig. 6.10. In practice, below a certain minimum, a retrogression in concrete strength, rather than the expected increase, takes place with a decrease in the W/C ratio. This reversed effect of the W/C ratio occurs because below this minimum the concrete is not workable enough to allow full compaction. Hence, under such conditions, voids remain in the concrete, and its strength is thereby reduced. That is, the W/C ratio law is valid, only if the concrete can be fully compacted.

Finally, air entrainment reduces concrete strength, and this effect should be allowed for in using the W/C ratio law. In this respect, it is usually assumed that the additional air content, ΔA, brought about by air entrainment, has the same effect on strength as the addition of an equivalent amount of water.

Hence, from strength considerations, the W/C ratio in air-entrained concrete is defined by

$$\omega = (W + \Delta A)/C \tag{6.5}$$

The air content in well-compacted concrete depends on aggregate size and generally varies from 10 to 30 litres/m^3. The air content in air-entrained concrete varies, similarly, from 45 to 65 litres/m^3. Hence, the additional air content, ΔA, in air-entrained concrete is usually 35 litres/m^3.

6.6. EFFECT OF TEMPERATURE

Temperature affects concrete strength through its effect on (i) the rate of hydration, (ii) the nature of concrete structure, and (iii) the rate of evaporation and the resulting drying out of the concrete. It may be noted that the preceding effects may be of a contradictory nature. Temperature, for example, accelerates hydration, and thereby the development of concrete strength. On the other hand, the increased rate of evaporation, associated with elevated temperatures, reduces the amount of water available, and thereby retards the rate of hydration and may even cause its complete cessation. Hence, in practice, the combined effect of temperature on strength varies and depends on the specific conditions considered.

It was explained earlier (see section 2.5.1) that the rate of cement hydration is considerably increased with the rise in temperature. As the strength of concrete depends on the porosity of the cement paste, and porosity, in turn, is determined by the degree of hydration, it is to be expected that the rate of strength development and concrete early-age strength will both increase with the rise in temperature as well. On the other hand, assuming that the effect of temperature on ultimate degree of hydration is small (see section 2.5.2), and provided the concrete is not allowed to dry, concrete later-age strength is not expected to be greatly temperature-dependent. That is, identical concretes, exposed to different temperatures, are expected to exhibit essentially the same later-age strength. It has been demonstrated, however, that while concrete cast and initially cured at high temperatures exhibits the expected increased early-age strength, its later-age strength is adversely affected [6.26–6.30], when, in this context, 'early-age' generally refers to ages up to 7 days and 'later-age' to ages over, say, 28 days. This effect of temperature is demonstrated in Fig. 6.11

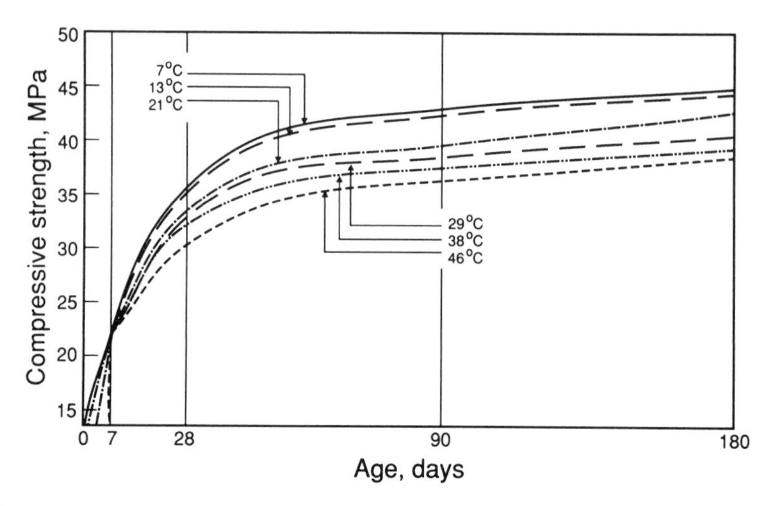

Fig. 6.11. Effect of initial curing temperature on concrete compressive strength. Specimens cast, sealed and maintained at the indicated temperature for 2 h, then stored at 21°C until tested. Type II cement, W/C ratio = 0·53. (Adapted from Ref. 6.26.)

and, generally, later-age strength reductions of some 25%, and more, were recorded when, during hydration, the concrete was exposed to elevated temperatures [6.27–6.30].

Although it is generally accepted that later-age strength of concrete is adversely affected by elevated temperatures, the relevant data available may differ considerably (Fig. 6.12). In the study summarised in part (A) of Fig. 6.12, later-age strength decreased gradually within the range of 5–46°C. On the other hand, within essentially the same range, an optimum temperature was observed which imparted to the concrete maximum strength (Fig. 6.12(B)). In this specific case, the optimum temperature of 13°C was observed in concretes made of types I and II cements, but in other studies optimum temperatures of 20–30°C (Fig. 6.12(D)) and 40°C (Fig. 6.13) were observed as well. Moreover, in yet another study, a critical temperature (i.e. a temperature at which the concrete achieves minimum strength), rather than an optimum temperature, was observed (Fig. 6.12(C)). The conflicting data reflect the complicated nature of the temperature effect, and support previous observation that several factors, rather than a single one, are involved. Apparently, these factors are affected differently by different test conditions, and their combined effect on strength, therefore, varies as well.

Some of the factors which may explain the adverse effect of elevated temperatures on concrete strength, have been considered earlier in the text (see

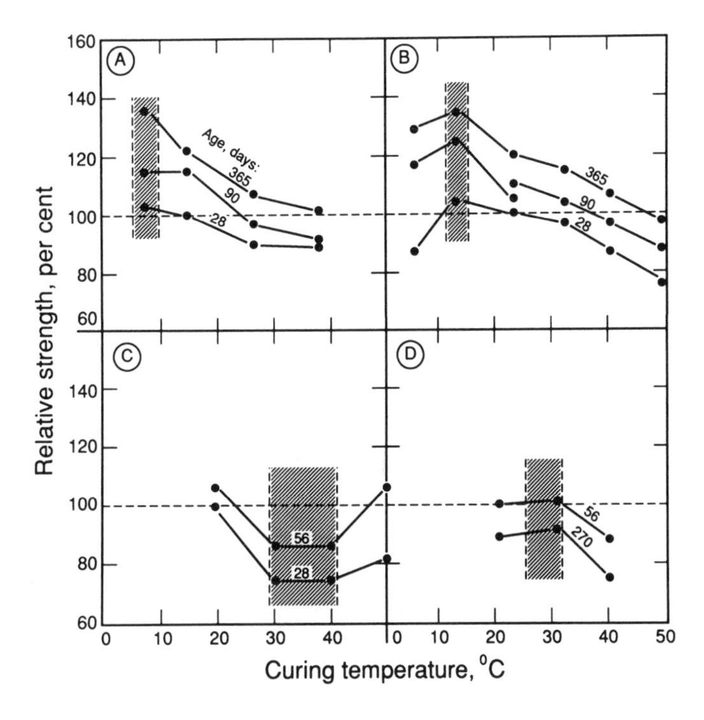

Fig. 6.12. Effect of initial curing temperature on later-age strength of concrete. (Adapted from the data in (A) Ref. 6.27, (B) Ref. 6.28 (C) Ref. 6.31, and (D) Ref. 6.32.)

sections 2.5.3 and 2.5.4). In this respect it was shown that this effect is attributable neither to changes in the composition of the hydration products nor to changes in the size of the gel particles. On the other hand, temperature was found to affect the nature of pore-size distribution, and a higher temperature was usually associated with a coarser system (Chapter 2, Fig. 2.12). It is to be expected from the failure mechanism of brittle materials, that a coarser pore system would result in a lower strength [6.33, 6.34]. Hence, the adverse effect of elevated temperatures can be explained, partly at least, by the coarser porosity which is brought about by such temperatures.

Another explanation is attributable to the effect of temperature on the optimum gypsum content of the cement. It was shown earlier (see section 1.3.1) that in practice gypsum is added in the amount required to give the cement the optimum SO_3 content. This optimum is determined for normal temperatures, whereas for elevated temperatures the optimum is significantly higher. As the optimum gives the concrete its maximum strength, any deviation from the latter would result in a lower strength. Such a deviation

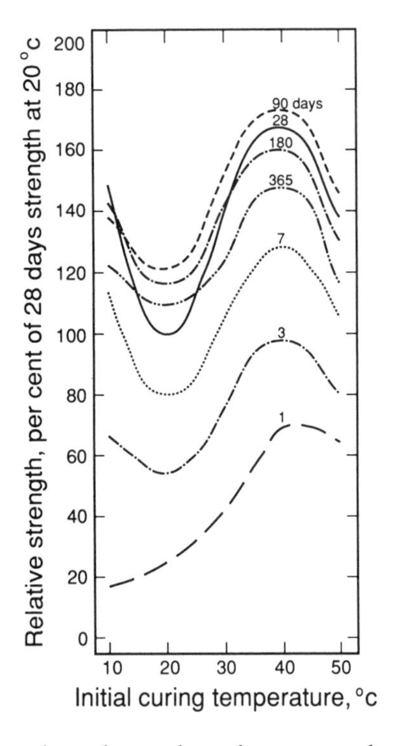

Fig. 6.13. Effect of initial curing temperature on concrete strength at different ages. Specimens cast and maintained covered at the indicated temperature for 20 h followed by 6 days curing in water at 21°C, and then stored at 21°C and 65% RH until tested. Ordinary Portland cement, W/C = 0·68. (Adapted from Ref. 6.29.)

takes place when the cement hydrates at a higher temperature, and thereby may partly explain the adverse effect of the latter on concrete strength.

In addition to the coarser pore system and the optimum gypsum content, the decrease in concrete strength with temperature may be attributed to the internal cracking and heterogeneity of the gel which may occur when hydration takes place under elevated temperatures. These possible temperature effects are discussed below.

6.6.1. Internal Cracking

In discussing plastic shrinkage (Chapter 5) it was explained that drying causes the fresh concrete to contract. When the contraction is restrained, tensile stresses are induced and the concrete cracks if, and when, the latter stresses exceed the tensile strength of the concrete at the time considered. It will be seen later (Chapter 7) that cracking may occur due to drying of the hardened concrete as well. Generally, however, when reference is made to cracking, it is usually meant to describe cracks which are detectable on the surface of the concrete. On the other hand, cracking may take place inside the concrete as well, due to the restraining effect of the aggregate particles. Such an internal

Fig. 6.14. Internal cracking in concrete exposed to drying at an early age. (Courtesy of C. Jaegermann, National Building Research Institute, Technion—Israel Institute of Technology, Haifa.)

cracking in a concrete specimen, which was exposed to intensive drying, is demonstrated in Fig. 6.14. It may be stipulated that the intensity of the internal cracking would increase with the severity of the drying conditions, and thereby explain the adverse effect of elevated temperatures on concrete strength.

It should be pointed out that the conclusion that exposing the concrete to intensive drying may result in internal cracking and strength reduction, is not valid if drying is limited to the very early age when the concrete is still plastic. In fact, when the concrete is allowed to dry during its plastic stage (i.e. up to 1–2 h after casting) its later-age strength may actually increase (Fig. 6.15). This increase in strength is attributed to the consolidation of the fresh concrete and the reduction in its effective W/C ratio which are brought about by the drying process. At this early stage the concrete is plastic enough to accommodate the associated contraction and, therefore, no internal cracking takes place. Longer drying, however, involves a brittle and weak concrete. Hence, cracking occurs to an extent which counteracts the beneficial effect of the earlier drying, and the net effect on concrete strength is negative. Under conditions relevant to Fig. 6.15, the negative effect became apparent when the early exposure of the concrete exceeded, say, 2 h.

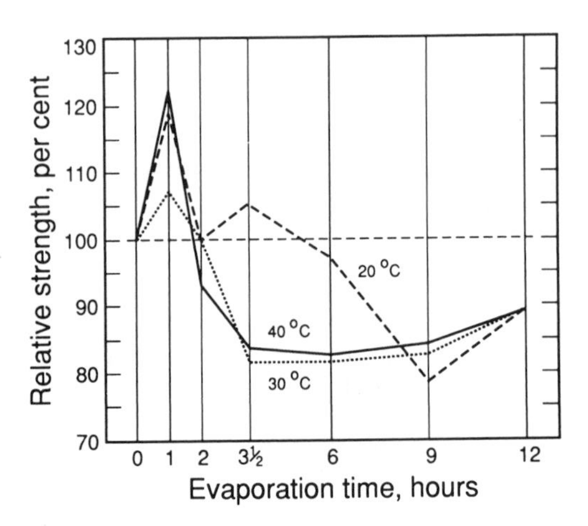

Fig. 6.15. Effect of early drying on 56 days strength of concrete containing 350 kg/m^3 OPC. Concrete exposed unprotected at the temperatures indicated. Wind velocity 20 km/h, RH 30%. (Adapted from Ref. 6.32.)

6.6.2. Heterogeneity of the Gel

It was suggested that the adverse effect of temperature on concrete strength is attributable to the heterogeneity of the gel which is brought about when the hydration takes place at elevated temperatures. It was stipulated that at low temperatures, when the hydration is relatively slow, there is ample time for the hydration products to diffuse and precipitate uniformly between the cement grains. On the other hand, when the cement hydrates at elevated temperatures, the high rate of hydration does not allow for such uniform precipitation to take place, and there is a tendency for the hydration products to precipitate in the immediate vicinity of the hydrating cement grains. Consequently, a highly concentrated and dense gel is formed around the hydrating cement grains, whereas a more porous, and therefore a weaker gel, is formed at a greater distance from the grains. This weaker part of the gel determines the strength of the set cement and its formation, therefore, may explain the detrimental effect of temperature on strength. This suggested mechanism of temperature effect is schematically described in Fig. 6.16.

It was also suggested that the formation of a comparatively dense layer around the cement grains retards further hydration to a greater extent than the less dense gel which is formed when the cement hydrates under normal temperatures (see section 2.4). That is, the adverse effect of temperature on concrete strength is attributated also to the lower ultimate degree of hydration which is reached when the hydration of the cement takes place under elevated

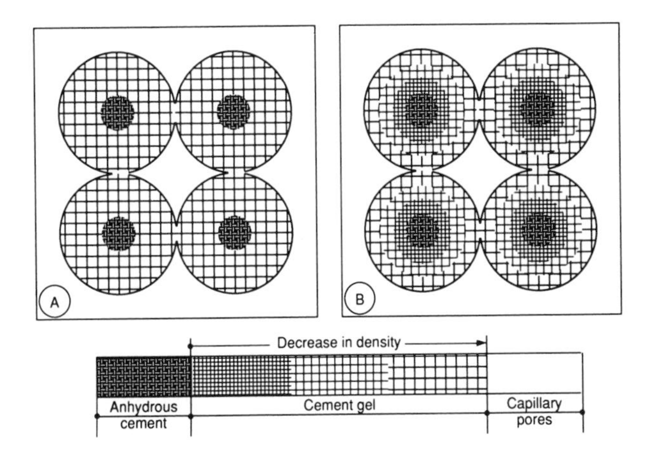

Fig. 6.16. Effect of temperature on the uniformity of the gel structure. (A) A gel of essentially the same density is formed at normal temperature, (B) a gel of non-uniform density is formed under elevated temperatures. Density is decreased with the distance from the hydrating cement grains. (Adapted from Ref. 6.35.)

temperatures. This latter conclusion may be questioned because such an effect of temperature on ultimate degree of hydration has not been observed in all cases (see section 2.5.2).

6.6.3. Type of Cement

It was pointed out earlier that the available data with respect to the effect of temperature on concrete strength are of a contradictory nature. Noting that concrete properties are very much related to the type of the cement used for its production, it is reasonable to assume that the sometimes contradictory nature of the temperature effect may be attributed to differences in the chemical composition and other properties of the cements involved. This relation, if any, between the cement composition and temperature effect on concrete strength is, obviously, of practical importance and if established would enable the selection of the most suitable cement for given climatic conditions. Relevant data, however, are limited and not conclusive. It may be concluded from Fig. 6.17, for example, that, strengthwise, the use of type V cement (i.e. sulphate-resisting cement) is preferable under hot, dry conditions, i.e. it can be seen that a reduction of some 10% was observed in concrete strength at the age of 360 days, when the latter cement was used, as compared to 20–26% when ordinary Portland cements were used. This conclusion,

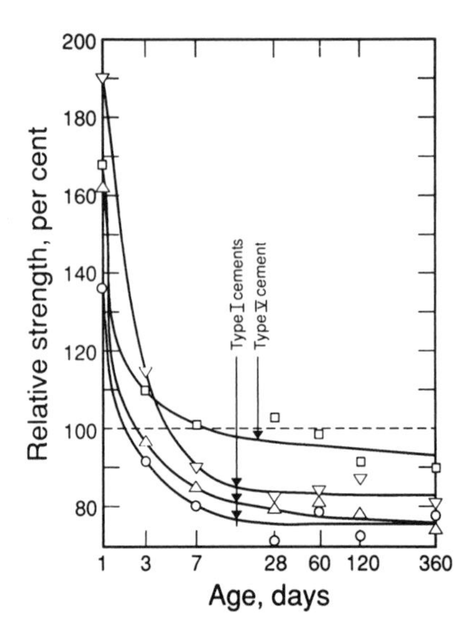

Fig. 6.17. Strength of concretes made with various Portland cements and cured for 24 h at 40°C and 45% RH, in relation to otherwise the same concretes cured at 20°C and 70% RH. (Adapted from Ref. 6.30.)

although supported by some data [6.36] does not agree with some other data [6.28, 6.29]. At present, no final conclusion can be reached, and this specific aspect requires further research.

6.7. SUMMARY AND CONCLUDING REMARKS

Concrete strength is determined by (i) the strength of the cement paste, (ii) the strength of the paste–aggregate bond, and (iii) some aggregate properties. In this respect, the strength of the paste is very significant, and all factors which affect the latter also affect concrete strength. Amongst these factors, the W/C ratio is most important, and under otherwise the same conditions, concrete strength is determined by this ratio alone. Accordingly, concrete strength, S, may generally be expressed by $S = A/B^{\omega}$, where ω is the W/C ratio, and A and B are constants which depend on the type of cement and aggregate involved, and on curing conditions, age, and testing method. This expression, sometimes referred to as 'Abrams' law', plays a very important role in concrete mix design.

The strength of the paste–aggregate bond depends on the strength of the paste (i.e. again on the W/C ratio), and on some properties of the aggregate. Generally, an improved bond, and consequently an improved strength, are to be expected with the increase in the roughness of the aggregate surface. The chemical and mineralogical composition of the aggregate may have some effect on bond strength, but this effect is usually not considered in daily practice.

Concrete strength decreases with the use of coarser aggregate, and increases with increase in aggregate concentration and rigidity, i.e. its modulus of elasticity. However, when ordinary aggregates are considered, all these effects of aggregate properties are not considered in everyday practice.

Generally, the preceding discussion is valid for compressive as well as for tensile strength of concrete. Quantitatively, however, some factors affect these two strengths differently. For example, the tensile strength is less sensitive to variations in the W/C ratio. Consequently, the ratio of tensile to compressive strength is not constant and decreases with increasing concrete strength. In most cases, it varies from 0·10 to 0·20 for strong and weak concretes, respectively, when the tensile strength is determined in flexure.

Temperature affects concrete strength through its effect on (i) the rate of hydration, (ii) the nature of concrete structure, and (iii) the rate of evaporation and the resulting drying out of the concrete. Generally, due to the increased rate of hydration, elevated temperatures increase early-age strength of concrete. Its later-age strength, however, is adversely affected. This adverse effect is attributable to a non-uniform gel of a coarser porosity which is produced under elevated temperatures, to internal cracking, and to a different optimum gypsum content which characterises such temperatures. Strength-wise, it is not yet clear which type of Portland cement, if any, is preferable under elevated temperatures.

REFERENCES

6.1. Soroka, I., *Portland Cement Paste and Concrete*. The Macmillan Press Ltd, London, UK, 1979, pp. 87–9.

6.2. Spooner, D.C., The stress–strain relationship for hardened cement pastes in compression. *Mag. Concrete Res.*, **24**(79) (1972) 85–92.

6.3. Feldman, R.F. & Beaudoin, J.J., Microstructure and strength of hydrated cement. In *Symp. Chem. of Cements*, Moscow, 1974.

6.4. Alexander, K.M., Wardlaw, J. & Gilbert, D.J., Aggregate cement bond, cement paste strength and the strength of concrete. In *Proc. Conf. Structure of Concrete and its Behaviour Under Load*. London, 1965. Cement and Concrete Association, London, 1968, pp. 59–92.

6.5. Hsu, T.T.C. & Slate, F.O., Tensile bond strength between aggregate and cement paste or mortar. *Proc. ACI*, **60**(4) (1963), 465–86.

6.6. Kaplan, M.F., Flexural and compressive strength of concrete as affected by the properties of the coarse aggregate. *Proc. ACI*, **55**(11) (1959), 1193–208.
6.7. Teychenne, D.C., Nicholls, J.C., Franklin, R.E. & Hobbs, D.W., *Design of Normal Concrete Mixes*. Dept. of Environment, British Research Establishment, Garston, Watford, UK, 1988.
6.8. Lyubimova, T.Yu. & Pinus, R.E., Crystallisation structure in contact zone between aggregate and cement in concrete. *Kolloidnyi Zhurnal*, **24**(5) (1962), 578–87 (in Russian).
6.9. Farran, J., Mineralogical contributions to the study of adhesion between the hydrated constituents of cement and embedded materials. *Rev. Mater. Constr. Trav.*, **430/1** (1956), 155–72; **492** (1956), 191–209 (in French).
6.10. Buck, A.L. & Dolch, W.L., Investigation of a reaction involving non-dolomitic limestone aggregate in concrete. *Proc. ACI*, **63**(7) (1966), 755–65.
6.11. Jarmontowicz, A. & Krzywoblocka-Laurow, R., Contact zone between calcareous aggregate and cement paste in concrete. In *Proc. RILEM Symp. on Aggregates and Fillers*, Budapest, Hungary, 1978, pp. 197–204.
6.12. Meyers, S., How temperature and moisture content may affect the durability of concrete. *Rock Products*, **54**(8) (1951), 153–7.
6.13. Mayer, F.M., The effect of different aggregates on the compressive strength and modulus of elasticity of normal concrete. *Beton*, **22**(2), (1972), 61–2 (in German).
6.14. Singh, B.G., Specific surface of aggregates related to compressive and flexural strength of concrete. *Proc. ACI*, **54**(10) (1958), 897–907.
6.15. Walker, S. & Bloem, L., Effects of aggregate size on concrete properties. *Proc. ACI*, **57**(3) (1960), 283–98.
6.16. Cordon, W.A. & Gillespie, H.A., Variables in concrete aggregates and Portland cement paste which influence the strength of concrete. *Proc. ACI*, **60**(8) (1963), 1029–52.
6.17. Hobbs, D.W., The compressive strength of concrete: A statistical approach to failure. *Mag. Concrete Res.*, **24**(80) (1972), 127–38.
6.18. Hobbs, D.W., The stress and deformation of concrete under short-term loading: A review. Cement and Concrete Association, Technical Report No. 42–484, London, UK, 1973.
6.19. Gilkey, H.J., Water cement ratio versus strength—another look. *Proc. ACI*, **57**(10) (1961), 1287–312.
6.20. Erntroy, H.C. & Shacklock, B.W., Design of high strength concrete mixes. In *Proc. Symp. on Mix Design and Quality Control of Concrete*, London, 1954; pp. 55–73.
6.21. McIntosh, J.D., Basic principles of concrete mix design. In *Proc. Symp. on Mix Design and Quality Control of Concrete*, Cement and Concrete Association, London, 1954, pp. 3–18.
6.22. Wright, P.J.F. & McCubin, A.D., The effect of aggregate type and aggregate

cement ratio on compressive strength of concrete. Road Research Note RN/ 1819, Road Research Laboratories, Crawthorne, UK, 1952.

6.23. Ishai, O., On the dual type fracture in hardened cement mortars. *Bull. Res. Council Israel*, **7C**(3) (1959), 147–54.

6.24. Ishai, O., Influence of sand concentration on deformation of mortar beams under low stress. *Proc. ACI*, **58**(5) (1961), 611–23.

6.25. Abrams, D.A., Design of concrete mixes. Bull. No. 1, Structure of Materials Research Laboratories, Lews Inst., Chicago, 1918. Reprinted in *A Selection of Historic American Papers on Concrete, 1926–1976* (ACI Spec. Publ. SP 52), ed. H. Newlon Jr. ACI, Detroit, MI, USA, 1976, pp. 309–30.

6.26. Price, W.H., Factors influencing concrete strength. *J. ACI*, **47**(5) (1951) 417–32.

6.27. US Bureau of Reclamation, Effect of initial curing temperatures on the compressive strength and durability of concrete. Concrete Laboratory Report No. C-625, US Dept. of Interior, Denver, CO, USA, July 29, 1952.

6.28. Klieger, P., Effect of mixing and curing temperature on concrete strength. *J. ACI*, **54**(12) (1958), 1063–81.

6.29. Soroka, I. & Peer, E., Influence of cement composition on compressive strength of concrete. In *Proc. RILEM 2nd Int. Symp. on Concrete and Reinforced Concrete in Hot Countries*, Haifa, 1971, vol. I, Building Research Station—Technion, Israel Institute of Technology, Haifa, pp. 243–58.

6.30. Shalon, R. & Ravina, D., The effect of elevated temperature on strength of portland cements. In *Temperature and Concrete* (ACI Spec. Publ. SP25), ACI, Detroit, MI, USA, 1970, pp. 275–89.

6.31. Shalon, R. & Ravina, D., Studies in concreting in hot countries. In *Proc. RILEM Intern. Symp. on Concrete and Reinforced Concrete in Hot Countries*, Haifa, 1960, Vol. 1. Building Research Station—Technion, Israel Institute of Technology, Haifa.

6.32. Jaegermann, C.H., Effect of exposure to high evaporation and elevated temperatures of fresh concrete on the shrinkage and creep characteristics of hardened concrete. DSc thesis, Faculty of Civil Engineering, Technion—Israel Institute of Technology, Haifa, Israel, July 1967 (in Hebrew with an English synopsis).

6.33. Griffith, A.A., The phenomena of rupture and flow in solids. *Phil. Trans. Roy. Soc.*, **A221**, (1920), 163–98.

6.34. Soroka, I., *Portland Cement Paste and Concrete*. The Macmillan Press Ltd, London, UK, pp. 76–81.

6.35. Verbeck, G.J. & Copeland, L.E., Some physical and chemical aspects of high pressure steam curing. In *Menzel Symp. on High Pressure Steam Curing*. (ACI Spec. Publ. SP32). ACI, Detroit, MI, USA, 1972, pp. 1–13.

6.36. Butt, Y.M., Kolbasov, V.M. & Timashev, V.V., High temperature curing of concrete under atmospheric pressure. In *Proc. Symp. on Chem. of Cement*. Tokyo, 1968, Part III, The Cement Association of Japan, Tokyo, pp. 437–71.

6.37. Wright, P.J.F., The design of concrete mixes on the basis of flexural strength. In *Proc. Symp. on Mix Design and Quality Control of Concrete*. London, 1954. Cement and Concrete Association, London, pp. 74–6.

Chapter 7

Drying Shrinkage

7.1. INTRODUCTION

It was explained earlier that hardened cement is characterised by a porous structure, with a minimum porosity of some 28%, which is reached when all the capillary pores become completely filled with the cement gel (see section 2.4). This may occur, theoretically at least, in a well-cured paste made with a water to cement (W/C) ratio of about 0·40 or less. Otherwise, the porosity of the paste is much higher due to incomplete hydration and the use of higher W/C ratios. In practice, and under normal conditions, this is usually the case, and a porosity in the order of some 50%, and more, is to be expected.

The moisture content of a porous solid, including that of the hardened cement, depends on environmental factors, such as relative humidity etc., and varies due to moisture exchange with the surroundings. The variations in moisture content, generally referred to as 'moisture movement', involve volume changes. More specifically, a decrease in moisture content (i.e. drying) involves volume decrease commonly known as 'drying shrinkage', or simply 'shrinkage'. Similarly, an increase in moisture content (i.e. absorption) involves a volume increase known as 'swelling'. In practice, the shrinkage aspect is rather important because it may cause cracking (see section 7.5), and thereby affect concrete performance and durability. Swelling, on the other hand, is hardly of any practical importance. Hence, the following discussion is mainly limited to the shrinkage aspect of the problem. In this respect it should be pointed out that, although shrinkage constitutes a bulk property, it is usually

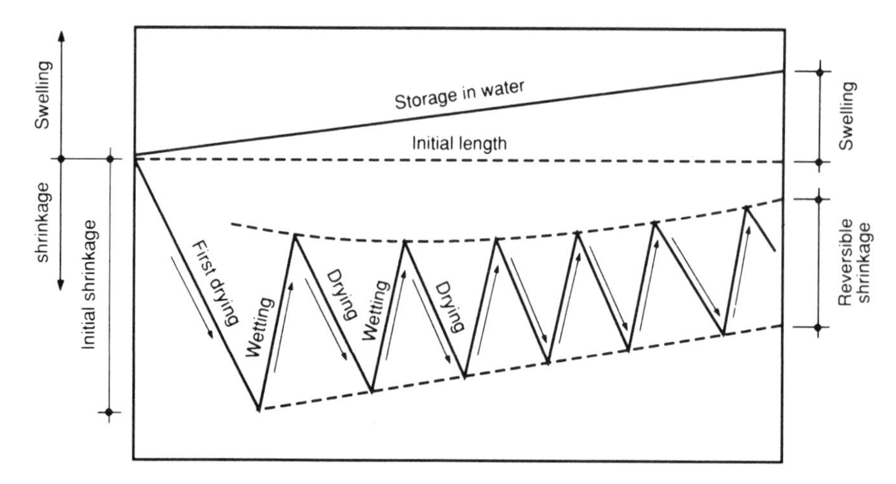

Fig. 7.1. Schematic description of volume changes in concrete exposed to alternate cycles of drying and wetting.

measured by the associated length changes and is expressed quantitatively by the corresponding linear strains, $\Delta l/l_0$.

7.2. THE PHENOMENA

A schematic description of volume changes in concrete, subjected to alternate cycles of drying and wetting, is given in Fig. 7.1. It may be noted that maximum shrinkage occurs on first drying, and a considerable part of this shrinkage is irreversible, i.e some part of the volume decrease is not recovered on subsequent wetting. Further cycles of drying and wetting result in additional, usually smaller, irreversible shrinkage. Ultimately, however, the process becomes more or less completely reversible. Hence, the distinction between 'reversible' and 'irreversible' shrinkage. In practice, however, such a distinction is hardly of any importance and the term 'shrinkage' usually refers to the maximum which occurs on first drying.

7.3. SHRINKAGE AND SWELLING MECHANISMS

As mentioned earlier, shrinkage is brought about by drying and the associated decrease in the moisture content in the hardened cement. A few mechanisms have been suggested to explain this phenomenon, and these are briefly discussed below. It may be noted that the following discussion mainly considers the cement paste. In principle, however, it is fully applicable to concrete because the presence of the aggregate in the paste hardly affects the shrinkage mechanism as such. On the other hand, the aggregate concentration and properties affect shrinkage quantitatively, but this aspect is dealt with later.

7.3.1. Capillary Tension

On drying, a meniscus is formed in the capillaries of the hardened cement and the formation of the meniscus brings about tensile stresses in the capillary water.

The tensile stresses in the capillary water must be balanced by compressive stresses in the surrounding solid. Hence, the formation of a meniscus on drying subjects the paste to compressive stresses which, in turn, cause elastic volume decrease. Accordingly, shrinkage is considered to be an elastic deformation. If this is indeed the case, it is to be expected that shrinkage will decrease, under otherwise the same conditions, with an increase in the rigidity of the solid, i.e. with an increase in its modulus of elasticity. In a cement paste the modulus of elasticity increases with strength which, in turn, is determined by the W/C ratio. That is, other things being equal, shrinkage is expected to decrease with a decrease in the W/C ratio or, alternatively, with an increase in strength. This is, indeed, the case which is further discussed in section 7.4.2.5.

It must be realised that the preceding mechanism of capillary tension is not complete because, contrary to experimental data and experience, it predicts the recovery of shrinkage at some later stage of the drying process. In practice, this is not the case and shrinkage occurs continuously as long as the drying of the paste takes place. Hence, it is usually assumed that the mechanism of capillary tension is significant mainly at the early stages of drying, i.e. when the relative humidity of the surroundings exceeds, say, 50%. It is further assumed that at lower humidities other mechanisms become operative, to such an extent that their effect is more than enough to compensate for the expected recovery due to the decrease in the capillary tension (see sections 7.3.2–7.3.4). Hence, the observed continued shrinkage on drying.

7.3.2. Surface Tension

Molecule A (Fig. 7.2), well inside a material, is equally attracted and repelled from all directions by the neighbouring molecules. This is not the case for

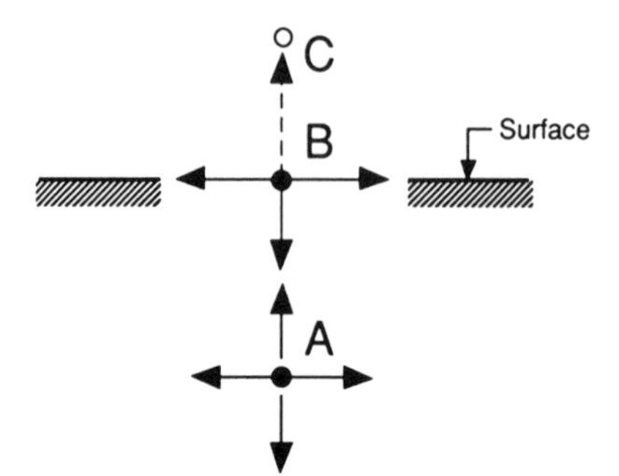

Fig. 7.2. Schematic representation of surface tension.

molecule B at the surface for which, because of lack of symmetry, a resultant force acts downwards at right angles to the surface. As a result, the surface tends to contract and behaves like a stretched elastic skin. The resulting tension in the surface is known as 'surface tension'.

The resultant force, acting downwards at right angles to the surface, induces compressive stress inside the material, and brings about elastic deformations. It can be shown that for spherical particles the induced stresses increase with an increase in surface tension and a decrease in the radius of the sphere. In colloidal-size particles, such as the cement gel particles, the induced stresses may be rather high and produce, therefore, detectable volume changes.

Changes in surface tension and the associated induced stresses, are brought about by changes in the amount of water adsorbed on the surface of the material, i.e. on the surface of the gel particles. It can be seen (Fig. 7.2) that an adsorbed water molecule, C, acts on molecule B in the opposite direction to the resultant force. The force, therefore, decreases, causing a corresponding decrease in surface tension. As a result, the compressive stress in the material is reduced and its volume increases due to elastic recovery, i.e. 'swelling' takes place. Similarly, drying increases surface tension and the increased compressive stress causes volume decrease, i.e. 'shrinkage' occurs. In other words, the proposed mechanism attributes volume changes to variations in surface tension of the gel particles which are brought about by variations in the amount of adsorbed water. It should be noted that only physically adsorbed water affects surface tension. Hence, the suggested mechanism is valid only at low humidities where variations in the water content of the paste are mainly due to variations in the amount of such water. At higher humidities, some of the water in the paste (i.e. capillary water) is outside the range of surface forces and a change in the amount of the so-called 'free' water does not affect surface tension. Accordingly, it has been suggested that the surface tension mechanism is only operative up to the relative humidity of 40% [7.1].

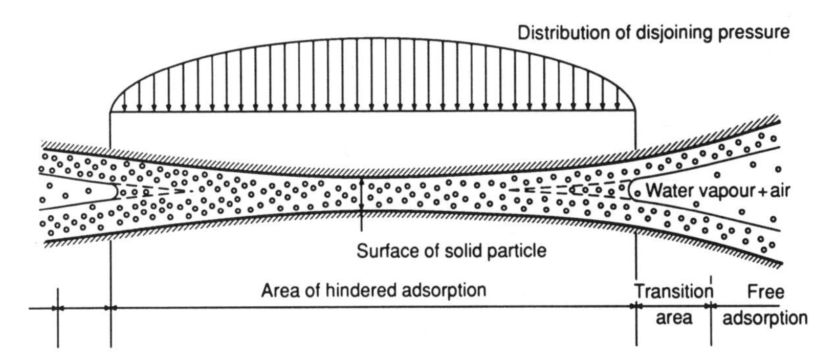

Fig. 7.3. Schematic description of areas of hindered adsorption and the development of swelling pressure. (Adapted from Ref. 7.2 in accordance with Power's model [7.3].)

7.3.3. Swelling Pressure

At a given temperature, the thickness of an adsorbed water layer on the surface of a solid is determined by the ambient relative humidity, and increases with an increase in the latter. On surfaces which are rather close to each other the adsorbed layer cannot be fully developed in accordance with the existing relative humidity. Such surfaces are sometimes referred to as 'areas of hindered adsorption'. In these areas a 'swelling' or 'disjointing' pressure develops and this pressure tends to separate the adjacent particles, and thereby cause swelling. This mechanism is schematically described in Fig. 7.3.

As mentioned earlier, the thickness of the adsorbed water layer increases with relative humidity and, in accordance with the preceding mechanism, the swelling pressure increases correspondingly. Hence, the swelling of the cement paste increases with an increase in its moisture content. A decrease in relative humidity causes drying. Consequently, the thickness of the adsorbed layer, and the associated swelling pressure, are decreased. When the swelling pressure is decreased, the distance between the mutually attracted gel particles is reduced, i.e. shrinkage takes place. In other words, according to this mechanism, volume changes are brought about by changes in interparticle separation which, in turn, are caused by variations in swelling pressure.

7.3.4. Movement of Interlayer Water

The calcium silicate hydrates of the cement gel (see section 2.4), are characterised by a layered structure. Hence, exit and re-entry of water in and out of such a structure, affect the spacing between the layers and thereby cause

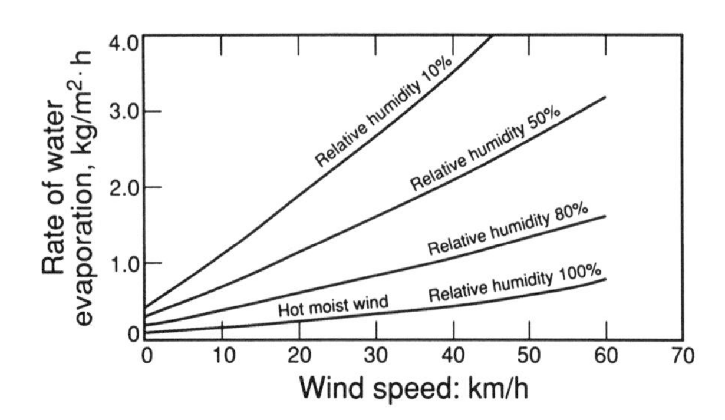

Fig. 7.4. Effect of wind velocity and relative humidity on the rate of water evaporation from concrete. Ambient and concrete temperature, 30°C. (Adapted from Ref. 7.5.)

volume changes. Accordingly, the exit of water on drying reduces the spacing and brings about a volume decrease, i.e. shrinkage. On the other hand, re-entry of water on rewetting increases the spacing, and thereby causes a volume increase, i.e. swelling [7.4].

7.4. FACTORS AFFECTING SHRINKAGE

As has been explained earlier, shrinkage is brought about by the drying of the cement paste. Consequently, all environmental factors which affect drying would affect shrinkage as well. Shrinkage is also affected by concrete composition and some of its properties. All of these factors which determine shrinkage are discussed below in some detail.

7.4.1. Environmental Factors

Environmental factors which affect drying include relative humidity, temperature and wind velocity. These effects are, of course, well known, and already have been discussed in section 5.2.1.1. The effect of the environmental factors is partly demonstrated again in Fig. 7.4 and considering the data of this figure, as well as the data of Fig. 5.4, it is clear that the intensity of the drying (i.e. the rate of evaporation) increases with the decrease in relative humidity and the increase in temperature and wind velocity. In other words, in a hot environment, and particularly in a hot, dry environment, both the rate and

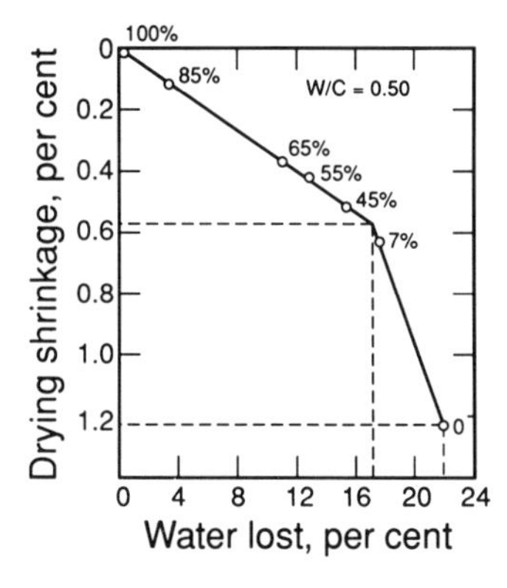

Fig. 7.5. Effect of water loss on shrinkage of cement paste. (Adapted from Ref. 7.6.)

amount of shrinkage are expected to be greater than under moderate climatic conditions. It will be seen later (see section 7.5) that shrinkage may cause cracking, and the possibility of such cracking is increased, the greater the shrinkage and the earlier it occurs. Hence, shrinkage-induced cracking must be considered a distinct possibility in a hot, dry climate, and suitable means are to be employed (i.e. adequate protection and curing) in order to reduce the risk of such cracking.

Considering the mechanisms which have been suggested to explain shrinkage (see section 7.3), it is evident that shrinkage is expected to increase with the increase in the intensity of drying, i.e. with the increase in the amount of water lost from the drying concrete. This is, indeed, the case, as is demonstrated, for example, by the data of Fig. 7.5. This relationship is not necessarily linear but, generally, it is characterised by two distinct stages. In the first stage, when drying takes place in the higher humidity region, a relatively large amount of water is lost but only a small shrinkage takes place. In the second stage, however, when drying takes place at lower humidities, a much smaller water loss is associated with a considerably greater shrinkage. Accordingly, for example, under the conditions relevant to Fig. 7.5, a water loss of approximately 17% in the high humidity region resulted in a shrinkage of some 0·6%, whereas an additional loss of only 6% in the lower region doubled the shrinkage to 1·2%.

That the mechanism of capillary tension described earlier (see section 7.3.1), may be used to explain why, at early stages of drying, the amount of water lost is large compared to the resulting shrinkage. At the early stages, water evaporates from the bigger pores (i.e. the capillary pores) accounting for the comparatively large amount of water lost. The resulting shrinkage, however, is small because of the relatively large diameter of the pores involved. At later

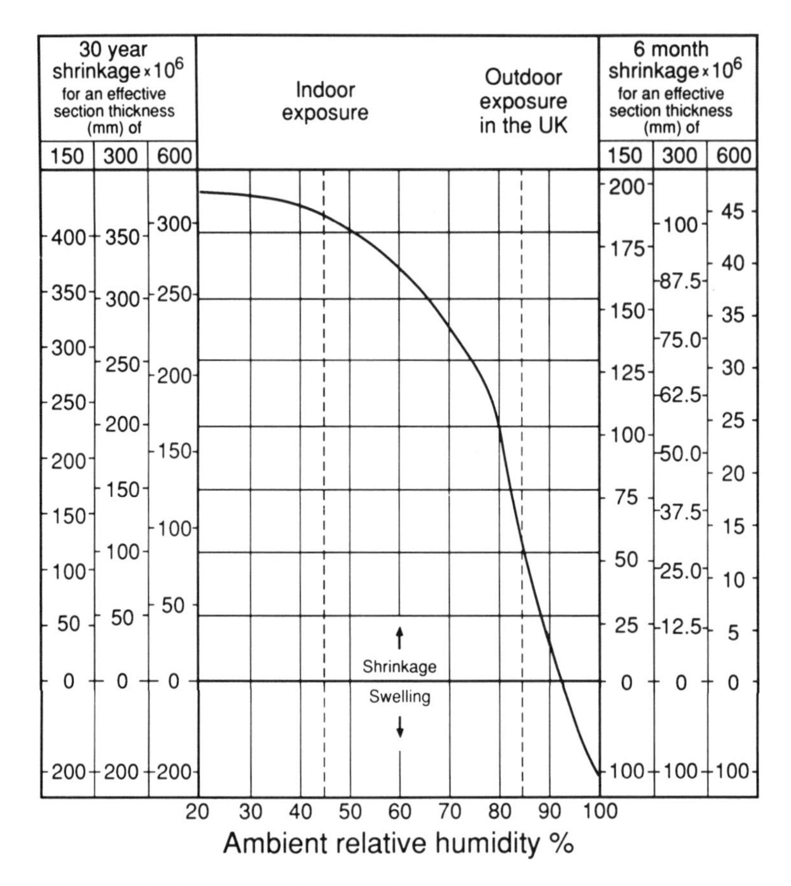

Fig. 7.6. Effect of relative humidity on shrinkage. (Adapted from Ref. 7.7.)

stages water evaporates from the smaller gel pores. Hence, the amount of water lost is comparatively small, but the shrinkage is relatively high.

The preceding conclusion that, due to a more intensive drying, a higher shrinkage is to be expected in hot, dry environment is well recognised and is reflected, for example, in estimating shrinkage with respect to ambient relative humidity in accordance with British Standard BS 8110, Part 2, 1985. It can be seen (Fig. 7.6) that, indeed, shrinkage is highly dependent on relative humidity and, for example, the decrease in the latter from 85 to 45% is expected to increase shrinkage approximately by a factor of three.

It was shown earlier (Chapter 6, Fig. 6.15) that short-time exposure (i.e. 1–2 h) of fresh concrete to intensive drying actually increased concrete later-age strength, but longer exposure periods caused strength reductions. It will be seen later (see section 7.4.2.5) that reduced shrinkage is to be expected in stronger concretes. Hence, short-time exposure of fresh concrete is expected

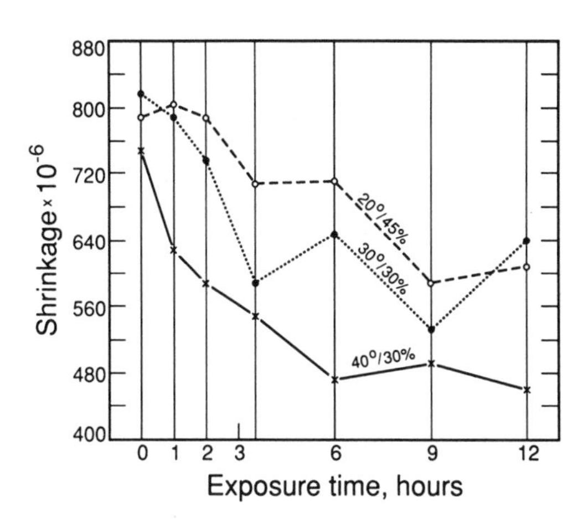

Fig. 7.7. Effect of early exposure at the temperatures and relative humidities indicated (wind velocity 20 km/h), on shrinkage of concrete containing 350 kg/m^3 ordinary Portland cement. Drying at 20°C and 50% RH from the age of 28 days to the age of 425 days. (Adapted from Refs 7.8 and 7.9.)

to reduce shrinkage and, indeed, such a reduction was observed in concrete which was exposed for a short time to intensive drying (Fig. 7.7). It can be seen, however, that while the beneficial effect of the early drying on concrete strength was limited to short exposure times of 1–2 h, its reducing effect on shrinkage was evident for exposure times as long as 6–9 h. This difference in exposure times is attributable to the effect of drying on the structure of the concrete which, in turn, affects differently strength and shrinkage. At a very early age, when the concrete is still plastic and can accommodate volume changes, drying causes consolidation of the fresh mix and reduces the effective W/C ratio. Hence, the increased strength and the associated reduced shrinkage. At a later age, however, setting takes place and the concrete cannot further accommodate volume changes, and internal cracking occurs (Chapter 6, Fig. 6.14). Such cracking reduces strength and more than counteracts the beneficial effect of the earlier drying. Hence, the net effect is a reduction in concrete strength. On the other hand, the presence of cracks, including internal cracks, reduces shrinkage because some of the induced strains are taken up by the cracks and are not reflected, therefore, in the bulk dimensions of the concrete. Hence, the reduction in measured shrinkage.

The reducing effect of early and short drying on shrinkage of concrete has also been observed by others under hot, dry (Fig. 7.8) and hot, humid (Fig. 7.9) environments. It must be realised, however, that this apparently beneficial effect of early drying has only very limited, if any, practical implication. The

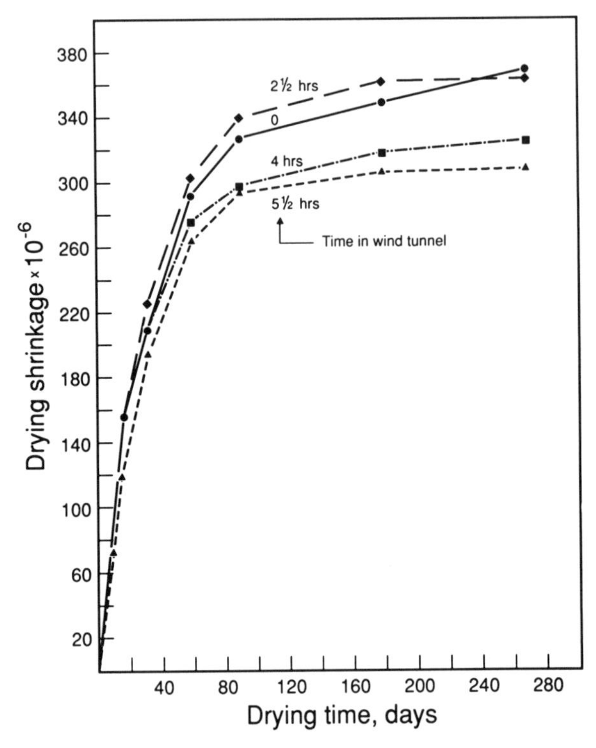

Fig. 7.8. Effect of early exposure in a wind tunnel, for the length of time indicated, on shrinkage of concrete stored from the age of 7 days at 20°C and 50% RH. (Adapted from Ref. 7.10.)

data in question were obtained from the laboratory testing of specimens, and in such specimens, contraction is only slightly restrained. This is, of course, not the case in practice where contraction is always restrained by the reinforcement, connection to adjacent members and friction. Consequently, under such conditions, the early exposure of concrete to drying, and particularly to intensive drying, is very likely to produce cracking and such exposure must, therefore, definitely be avoided. In fact, fresh concrete should be protected from drying as early as possible, and particularly in a hot, dry environment. Further discussion of this aspect is presented in Chapter 5.

7.4.2. Concrete Composition and Properties

7.4.2.1. Aggregate Concentration

In considering shrinkage, concrete may be regarded as a two-phase material consisting of cement paste and aggregates. Shrinkage of the cement paste,

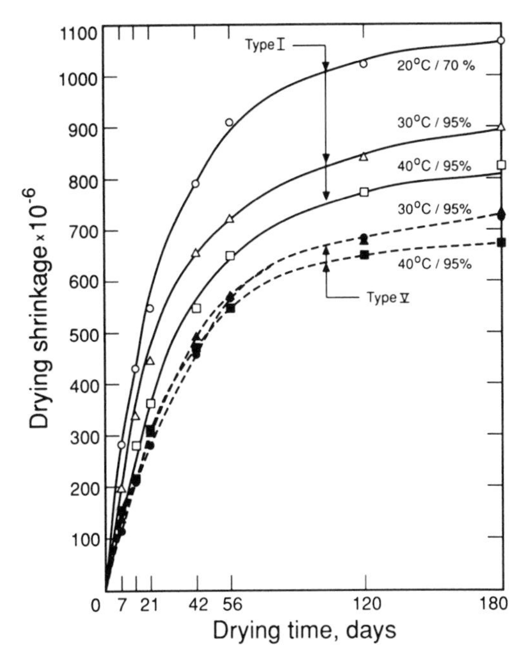

Fig. 7.9. Effect of 24 h exposure, at the temperatures and relative humidities indicated (still air), on shrinkage of concrete from the age of 28 days at 20°C and 50% RH. (Adapted from Ref. 7.11.)

when measured from the associated length changes, may reach some 0·5%, whereas that of normal concrete aggregates is much smaller. Hence, shrinkage of concrete is essentially determined by the shrinkage of the paste and its concentration in the concrete. That is, shrinkage of concrete is expected to increase with an increase in the paste content or, alternatively, with a decrease in that of the aggregate. This effect of aggregate concentration is supported by the experimental data presented in Fig. 7.10 and, indeed by the data of some others [7.12].

7.4.2.2. Rigidity of Aggregate

Noting that shrinkage of normal aggregates is very small compared to that of a cement paste, it follows that the presence of the aggregates would restrain shrinkage to an extent which depends on their rigidity. In discussing shrinkage mechanisms (see section 7.3), it was pointed out that shrinkage is actually an elastic deformation. Consequently, shrinkage is expected to depend on the concrete modulus of elasticity and to decrease with an increase in the latter, and vice versa. The rigidity of the aggregate affects concrete modulus of

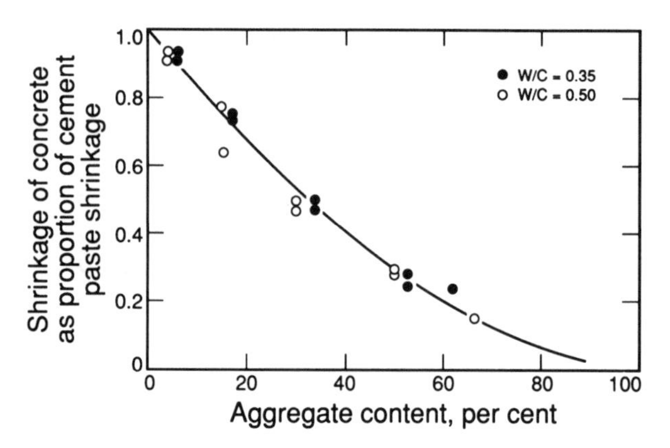

Fig. 7.10. Effect of aggregate concentration on shrinkage of concrete. (Adapted from Ref. 7.13.)

elasticity and thereby affects its shrinkage. That is, under otherwise the same conditions, a lower shrinkage is to be expected in a concrete made with a rigid aggregate than in a concrete made with a soft aggregate, such as lightweight aggregate. This latter conclusion, and the dependence of shrinkage on the modulus of elasticity of the concrete, are indicated by the data presented in Fig. 7.11.

The combined effect of aggregate concentration and rigidity on shrinkage of concrete can be expressed by the following formula:

$$S_c = S_p(1 - V_a)^n \tag{7.1}$$

where S_c and S_p are the corresponding shrinkage strains of the concrete and the paste, respectively; V_a is the aggregate concentration; and n represents the elastic properties of the aggregate.

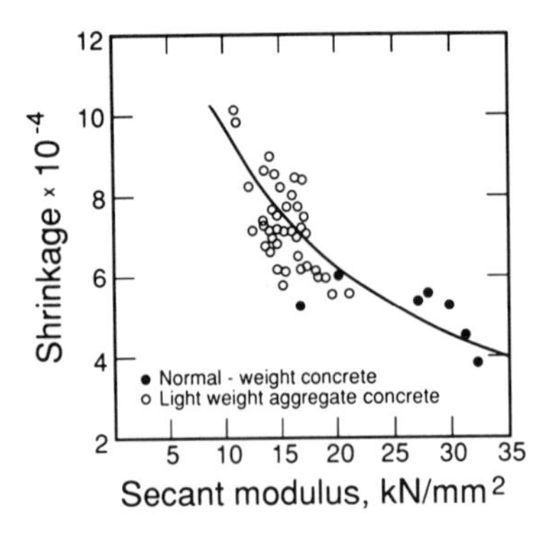

Fig. 7.11. The relation between shrinkage and concrete modulus of elasticity. (Adapted from Ref. 7.14.)

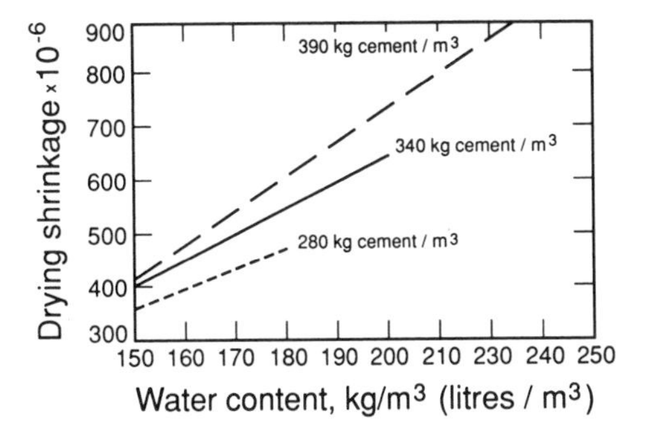

Fig. 7.12. Effect of water content on shrinkage of concrete made of different cement contents. (Adapted by Shirley [7.15] from Ref. 7.16.)

7.4.2.3. Cement Content

The paste concentration in concrete is determined by the cement content and increases with the increase in the latter. Hence, a greater concentration, and consequently greater shrinkage, is to be expected in cement-rich concrete than in lean concrete. This behaviour is demonstrated in Fig. 7.12.

7.4.2.4. Water Content

Shrinkage is related to the amount of water lost on drying (Fig. 7.5). Hence, under otherwise the same conditions, the more water that is lost, the higher the shrinkage to be expected. Accordingly, it may be argued that the higher the water content, the more water is available for drying and, consequently, a greater shrinkage will take place. That is, shrinkage is expected to increase with an increase in the water content. This conclusion is supported, for example, by the data of Fig. 7.12.

7.4.2.5. W/C Ratio

Shrinkage being an elastic deformation, is related to the modulus of elasticity of the concrete, and is expected to increase with the increase in the latter, and vice versa. The modulus of elasticity is related to concrete strength which, in turn, is determined by the W/C ratio. Hence, the W/C ratio is expected to affect shrinkage in a similar way, and a higher W/C ratio would result in a greater shrinkage. This effect is clearly indicated in Fig. 7.13.

Following the preceding discussion, it may be concluded that in order to produce concrete with a minimum shrinkage, the water and cement contents, as well as the W/C ratio, should be kept to a minimum. It may be noted, however, that the three are interrelated and selecting the value of any two determines the value of the third one. In practice, the water content is selected to give to the fresh concrete the required consistency, and the W/C ratio to give to the hardened concrete the required quality and durability. Hence, in

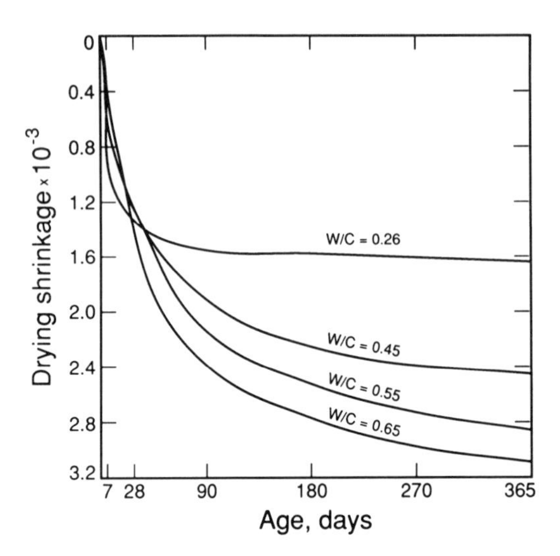

Fig. 7.13. Effect of W/C ratio on shrinkage of cement paste. (Adapted from Ref. 7.17.)

practice the cement content is determined in accordance with the pre-selected water content and W/C ratio.

Finally, it may be further noted that for a given cement content, increasing the water content is associated with a higher W/C ratio. Hence, it may be argued that the resulting increased shrinkage is not attributable to the increased water content *per se*, but to the associated higher W/C ratio. On the other hand, for the same water content increasing the cement content results in a lower W/C ratio. That is, in this case, shrinkage is determined by two opposing effects, i.e. the increased cement content is expected to increase shrinkage, whereas the reduced W/C ratio is expected to reduce it. In practice, however, cement-rich concrete usually exhibits a higher shrinkage than its lean counterpart.

7.4.2.6. Mineral Admixtures

The effect of admixtures on the drying shrinkage of concrete, due to the variation in the properties of the various types available, can only be discussed in a general way. Nevertheless, generally, the structure of the paste made from blended cements is usually characterised by a finer pore structure see (Chapter 3, Figs 3.3 and 3.15) and sometimes also by a lower porosity (Chapter 3, Fig. 3.14). On the other hand, considering shrinkage mechanisms (see section 7.3), a finer and a higher porosity would be associated with a higher shrinkage. Hence, it is to be expected that the shrinkage of concrete made of a blended

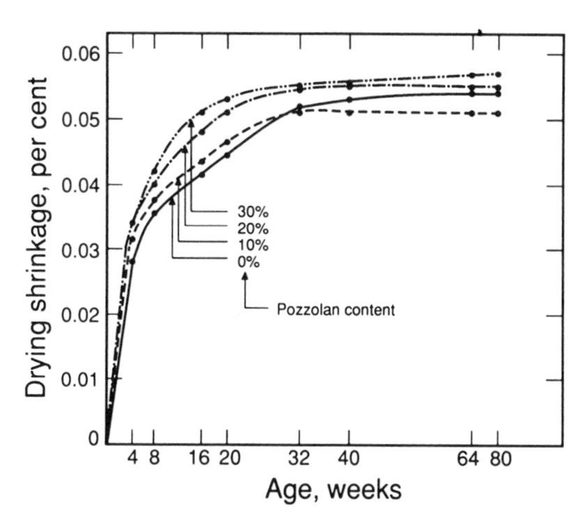

Fig. 7.14. Effect of replacing OPC with natural pozzolan (Santorin earth) on drying shrinkage of concrete. (Adapted from Ref. 7.19.)

cement will be higher than the shrinkage of its otherwise the same counterpart made of ordinary Portland cement (OPC). This expected effect is not necessarily supported by experimental data as can be seen from Figs. 7.14–7.17.

In considering the data of Figs 7.14–7.17 it may be noted that only when the cement was replaced with granulated blast-furnace slag the drying shrinkage of the concrete was clearly increased (Fig. 7.17). The corresponding increase, however, was rather small when the cement was replaced by natural pozzolan (Fig. 7.14) or fly-ash (Fig. 7.15) and disappeared completely when condensed silica fume was used for replacement (Fig. 7.16). Moreover, there exist also some contradictory data which indicate, for example, that the drying shrinkage of fly-ash concrete is actually lower than that of OPC concrete [7.18].

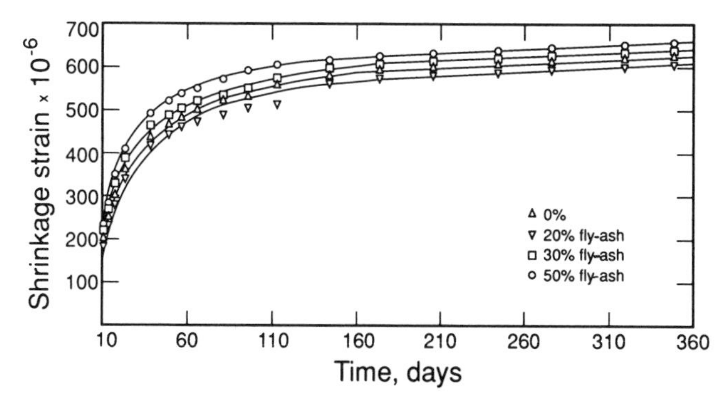

Fig. 7.15. Effect of replacing OPC with high-calcium fly-ash on drying shrinkage of concrete. (Adapted from Ref. 7.20.)

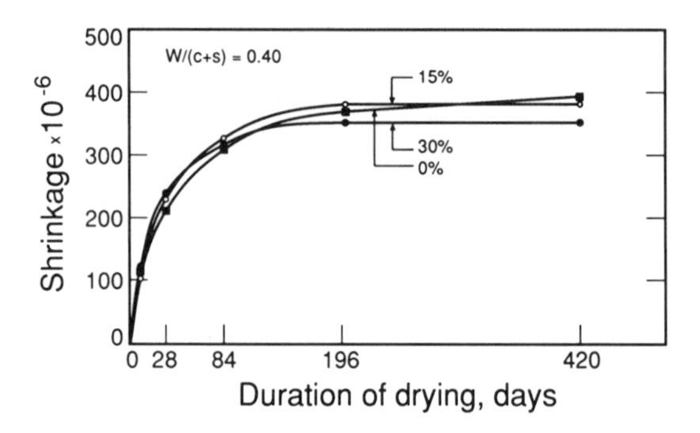

Fig. 7.16. Effect of replacing OPC with silica fume on drying shrinkage of concrete. (Adapted from Ref. 7.21.)

These contradictory data may be attributed to differences in testing conditions. In most cases, for example, shrinkage is compared in concretes of the same consistency, i.e. in concretes which may vary with respect to their water content and W/C ratio. As both water content and W/C ratio affect shrinkage test data may be affected differently and this may explain the sometimes contradictory nature of the results.

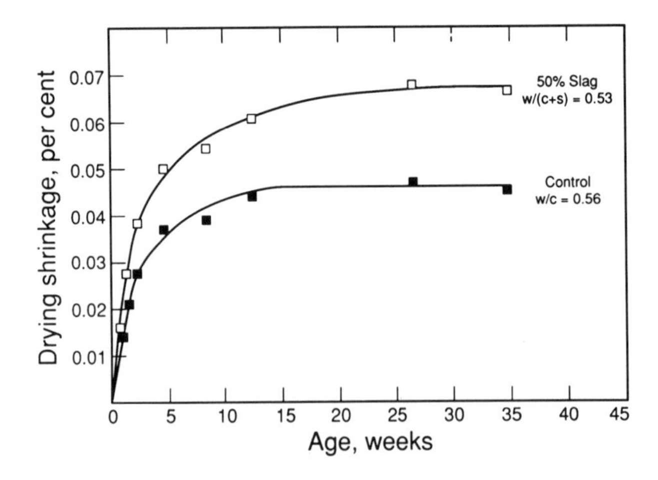

Fig. 7.17. Effect of replacing OPC with ground granulated blast-furnace slag on drying shrinkage of concrete. (Adapted from Ref. 7.22.)

7.5. SHRINKAGE CRACKING

Under normal conditions, shrinkage is unavoidable because, at one stage or another, drying of the concrete takes place. Moreover, in engineering applications shrinkage is restrained due to friction or rigid connections to adjacent concrete members. The restraining effect induces tensile stresses and the concrete may crack if, and when, the induced stresses exceed its tensile strength. The induced stress level depends on the intensity of drying, the degree of restraint, etc. Assuming elastic behaviour, the tensile stress, σ, is given by the expression $\sigma = \varepsilon_s E$, where ε_s is the shrinkage strain and E is concrete modulus of elasticity. Considering complete restraint, for which $\varepsilon_s = 200 \times 10^{-6}$, and $E = 30\,000\,\text{kN/mm}^2$, the induced tensile stress in the concrete will be $30\,000 \times 200 \times 10^{-6} = 6\,\text{N/mm}^2$, i.e. a stress level which is likely to exceed the concrete tensile strength, and particularly if it occurs at an early age when the concrete is comparatively weak. In practice, the stress level is usually lower than the preceding calculated level because creep effects relieve some of the tensile stress (Chapter 8), and because shrinkage is not fully restrained. Nevertheless, even under such conditions the resulting stress level may be high enough to cause cracking.

It should be stressed again that the likelihood of cracking is much greater in a hot environment, and particularly in a hot, dry environment, than in a moderate one because of the more intensive and rapid drying. Hence, under such conditions, cracking must be considered a distinct possibility and suitable means must be always employed in order to reduce this possibility, and perhaps even to eliminate it altogether. These include means to produce concrete with the lowest possible shrinkage, on the one hand, and means to delay its occurrence as long as possible, on the other.

(1) The concrete mix should be designed with the lowest possible water and cement contents. In this respect the use of coarser aggregates (i.e. aggregates of greater maximum particle size) with a low fines content, is to be preferred. The use of water-reducing admixtures should be favourably considered.

(2) The onset of shrinkage must be delayed as long as practically possible because strength development with time reduces the likelihood of cracking. That is, concrete should be protected from drying as early as possible, and as long as possible. In this respect it should be noted that this beneficial effect of the increased strength on cracking possibility is

somewhat reduced by the associated increase in the modulus of elasticity of the concrete. The induced stresses are equal to $\varepsilon_s E$. Hence, the increase in modulus of elasticity involves a higher stress level, and thereby counteracts the beneficial effect of the increased strength. In general, the modulus of elasticity, E, is related to concrete compressive strength, S, by the expression $E = k\sqrt{S}$, where k is a constant which depends on the specific conditions such as shape and size of test specimen. Accordingly, for example, a four-fold increase in strength involves only a two-fold increase in the modulus of elasticity. That is, the beneficial effect of the increased strength on shrinkage outweighs the opposing effect of the corresponding increase in the modulus of elasticity.

(3) Joints should be provided in concrete members in order to reduce the restraining effect of the structure on shrinkage. It is self-evident that the smaller the restraint, the lower the induced tensile stresses, and thereby the possibility of cracking is reduced.

It is generally accepted that the exact composition of the cement hardly affects shrinkage except when the gypsum content deviates significantly from the optimum (see section 1.3.1). In this respect, however, the possible use of 'shrinkage compensating' cements is sometimes suggested. The setting of such cements is accompanied by expansion, which when restrained by the reinforcement, induces compressive stresses in the hardened concrete [7.23]. These stresses compensate, partly or wholly, for the shrinkage induced tensile stresses, and thereby may prevent cracking. At present the use of such cements is limited and information on their performance in hot environments is hardly available. Hence, the possible use of such cements is not considered here.

7.6. SUMMARY AND CONCLUDING REMARKS

Variations in moisture content of the hardened cement paste are associated with volume changes. The decrease in the volume of the paste on drying is referred to as 'drying shrinkage' or simply 'shrinkage', and its increase on rewetting as 'swelling'. Shrinkage is related to water loss and, accordingly, all factors which affect drying such as relative humidity, temperature and air movement, affect shrinkage as well.

A few mechanisms have been suggested to explain shrinkage, namely, the mechanisms of capillary tension, surface tension, swelling pressure and movement of interlayer water. Shrinkage is actually an elastic deformation and is related, therefore, to concrete strength or, alternatively, to the W/C ratio. Shrinkage increases with an increase in the water and the cement contents and decreases with the rigidity of the aggregate and its content. The use of mineral admixtures or blended cements may be associated with a greater shrinkage.

In practice, shrinkage is restrained, and this restraint produces tensile stresses in the concrete. Consequently, concrete may crack if, and when, induced stresses exceed the tensile strength of the concrete. The possibility of cracking is enhanced in hot environments, and particularly in hot, dry environments, due to a more intensive and rapid drying. In order to reduce this possibility, concrete should be made with the lowest possible water and cement contents, and protected from drying as early, and as long as possible, after being placed. As well, joints should be provided in the concrete members in order to reduce the restraining effect of the structure.

REFERENCES

7.1 Wittmann, F.H., Surface tension, shrinkage and strength of hardened cement paste. *Mater. Struct.*, **1**(6) (1968), 547–52.

7.2 Bazant, Z.P., Delayed thermal dilatation of cement paste and concrete due to mass transport. *Nuclear Engng Design*, **14** (1970), 308–18.

7.3 Powers, T.C., Mechanism of shrinkage and reversible creep of hardened cement paste. In *Proc. Conf. Structure of Concrete and Its Behaviour Under Load*, London, 1965, Cement and Concrete Association, London, 1968, pp. 319–44.

7.4 Feldman, R.F. & Sereda, P.J., A new model for hydrated cement and its practical implications. *Engng J.*, **53** (1970), 53–9.

7.5 Egan, D.E., Concreting in hot weather. Notes on Current Practices, Note No. 15, Cement and Concrete Association of Australia, March 1984, pp. 7–10.

7.6 Verbeck, G.J. & Helmuth, R.A., Structure and physical properties of cement paste. In *Proc. Symp. Chem. of Cement*, Tokyo, 1968, Vol. 3, The Cement Association of Japan, Tokyo. pp. 1–37.

7.7 BS 8110, Structural use of concrete. Part 2: 1985—Code of practice for special circumstances. HMSO, London, UK.

7.8 Jaegermann, C.H., Effect of exposure to high evaporation and elevated temperatures of fresh concrete on the shrinkage and creep characteristics of hardened concrete. DSc thesis, Faculty of Civil Engineering, Technion—Israel

Institute of Technology, Haifa, Israel, July 1967 (in Hebrew with an English synopsis).

7.9 Jaegermann, C.H. & Glucklich, J., Effect of high evaporation during and shortly after casting on the creep behaviour of hardened concrete. *Mater. Struct.*, **2**(7) (1969), 59–70.

7.10 Jaegermann, C.H. & Traubici, M., Effect of heat curing by means of hot air blowers of concrete precast slabs. Report to the Ministry of Housing, Technion —Building Research Station, Haifa, Israel, 1978 (in Hebrew).

7.11 Shalon, R. & Berhane, Z., Shrinkage and creep of mortar and concrete as affected by hot-humid environment. In *Proc. RILEM 2nd Int. Symp. on Concrete and Reinforced Concrete in Hot Countries*, Haifa, 1971, Vol. I. Building Research Station—Technion, Israel, Institute of Technology, Haifa, pp. 309–21.

7.12 Powers, T.C., Fundamental aspects of concrete shrinkage. *Rev. Mater. Constr.*, **545** (1961), 79–85 (in French).

7.13 Pickett G., Effect of aggregate on shrinkage and a hypothesis concerning shrinkage. *Proc. ACI*, **52**(5) (1956), 581–90.

7.14 Richard, T.W., *Creep and Drying Shrinkage of Lightweight and Normal Weight Concrete.* Monograph No. 74, National Bureau of Standards, Washington, DC, USA, 1964.

7.15 Shirley, D.E., *Concreting in Hot Countries* (3rd edn). Cement and Concrete Association, Wexham Springs, Slough, UK, 1978 (reprinted 1985).

7.16 US Bureau of Reclamation, *Concrete Manual* (8th edn). Denver, CO, USA, 1975, p. 16, Fig. 8.

7.17 Haller, P., *Shrinkage and Creep of Mortar and Concrete.* Diskussionbericht No. 124, EMPA, Zurich, Switzerland, 1940 (in German).

7.18 Yamato, T. & Sugita, H., Shrinkage and creep of mass concrete containing fly ash. In *Fly Ash, Silica Fume, Slag and Other Mineral By-Products* (ACI Spec. Publ., SP 79, Vol. 1), ed. V.M. Malhotra. ACI, Detroit, MI, USA, 1983, pp. 87–102.

7.19 Mehta, P.K., Studies on blended Portland cements containing Santorin earth. *Cement Concrete Res.*, **11**(4) (1981), 507–18.

7.20 Yuan R.L. & Cook, J.E., Study of a class C fly ash concrete. In *Fly Ash, Silica Fume, Slag and Other Mineral By-Products* (ACI Spec. Publ. SP 79, Vol. I), ed. V.M. Malhotra. ACI, Detroit, MI, USA, 1983, pp. 307–19.

7.21 ACI Committee 226, Silica fume in concrete. *ACI Mater J.*, **84**(1) (1987), 158–66.

7.22 Hogan, F.J. & Meusel, J.W., Evaluation for durability and strength development of a ground granulated blast furnace slag. *Cement Concrete and Aggregates*, **3**(1) (1981), 40–52.

7.23 ACI Committee 223–83, Shrinkage compensating concrete. In *ACI Manual of Concrete Practice* (Part 1). ACI, Detroit, MI, USA, 1990.

Chapter 8
Creep

8.1. INTRODUCTION

Creep may be defined as the increase in deformation with time, excluding shrinkage, under a sustained stress. Such a deformation occurs in metals at elevated temperatures but in concrete it takes place at room temperatures as well. Hence, the importance of the creep behaviour in daily practice.

In the following discussion a distinction is not always made between cement paste and concrete. Creep behaviour of concrete is essentially similar to that of the paste because the aggregate hardly exhibits any creep. Hence, in discussing creep qualitatively, paste and concrete are interchangeable. On the other hand, aggregate properties and concentration affect creep quantitatively, and in this context there is a significant difference between creep of the cement paste and that of concrete. This aspect, however, is dealt with later in the text (see section 8.4.2.1).

Finally, creep is usually measured by the length changes involved and is expressed quantitatively by the corresponding strains, $\Delta 1/1_0$, or by the corresponding strains per unit stress. The latter is known as 'specific creep' (see section 8.4.2.2).

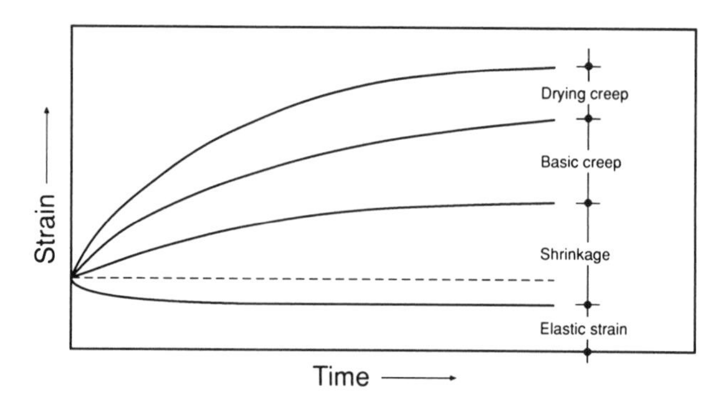

Fig. 8.1. Schematic description of the deformation with time of concrete under sustained compressive load and undergoing a simultaneous drying shrinkage.

8.2. THE PHENOMENA

On loading concrete undergoes an instantaneous deformation which is generally regarded as elastic, i.e. a deformation which appears and disappears completely immediately on application and removal of the load, respectively. If the load is sustained, the deformation increases, at a gradually decreasing rate, and may reach a value which is two to three times greater than the elastic deformation. If the concrete is allowed to dry when under load, shrinkage occurs simultaneously. Accordingly, creep is the increase in deformation with time under a sustained load excluding drying shrinkage. This is demonstrated in Fig. 8.1 for a concrete loaded in compression. It may be noted that the elastic deformation, contrary to creep and shrinkage, decreases with time. This is due to the increase in the modulus of elasticity which is associated with the increase in concrete strength.

Generally, the simultaneous drying of concrete is associated with increased creep (see section 8.4.1). Hence, a distinction is sometimes made between 'basic creep' and 'drying creep'. Basic creep is the creep which takes place when the concrete is in hygral equilibrium with its surroundings and, consequently, no simultaneous drying is involved. Accordingly, drying creep is the additional creep which is brought about by the simultaneous drying (Fig. 8.1). In most engineering applications the distinction between basic and drying creep is not important, and the term 'creep' usually refers to total creep, i.e. to the sum of basic and drying creep.

Similarly to shrinkage, creep is partly irrecoverable. On unloading, the strain decreases immediately due to elastic recovery. The instantaneous

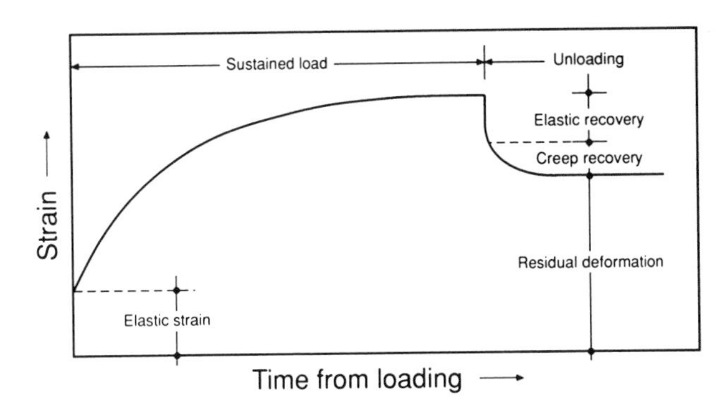

Fig. 8.2. Schematic description of creep and creep recovery in concrete in hygral equilibrium with its surroundings.

recovery is followed by a gradual decrease in strain which is known as 'creep recovery'. Creep recovery is not complete, approaching a limiting value with time. The remaining residual strain is the 'irreversible creep' (Fig. 8.2).

8.3. CREEP MECHANISMS

A few mechanisms have been suggested to explain creep of the cement paste and some of them are briefly presented here. It will be seen later that both creep and shrinkage are essentially affected the same way by the same factors and, indeed, to some extent, the two may be looked upon as similar phenomena. Consequently, some of the mechanisms which have been suggested to explain creep are actually an extension of the same mechanisms which have been suggested to explain shrinkage.

8.3.1. Swelling Pressure

In a previous discussion (see section 7.3.3), volume changes in the cement paste, due to variations in its moisture content, were attributed to variations in the swelling pressure brought about by variations in ambient relative humidity. It has been suggested that the same mechanism, induced by external loading, rather than the ambient humidity, may explain the reversible part of creep [8.1, 8.2]. That is, due to external loading some of the water between adjacent gel particles, i.e. some of the load-bearing water in areas of hindered adsorption

(Chapter 7, Fig. 7.3), is squeezed out into bigger pores (areas of unhindered absorption) by a time-dependent diffusion process. Consequently, the swelling pressure gradually decreases, the spacing between the gel particles is reduced and the volume of the paste is thereby decreased, i.e. creep takes place. When the paste is unloaded, the pressure on the load-bearing water is relieved, and a reversed process takes place. That is, the water gradually diffuses back from the areas of unhindered absorption, and the swelling pressure gradually increases to the level determined by the ambient relative humidity. This resulting increase in the swelling pressure causes a volume increase, i.e. creep recovery is taking place.

8.3.2. Stress Redistribution

On application, the external load is distributed between the liquid and the solid phases of the concrete. Under sustained loading the compressed water diffuses from high to low pressure areas and, consequently, a gradual transfer of the load from the water to the solid phase takes place. Hence, the stress in the solid gradually increases causing, in turn, a gradual volume decrease, i.e. creep. That is, creep may be regarded as a delayed elastic deformation [8.3, 8.4]. Accordingly, a lower creep is to be expected in a stronger concrete because such a concrete has a higher modulus of elasticity. Similarly, a higher creep is to be expected at a higher moisture content, because the higher the moisture content the greater the part of the load which is initially taken by the water and later transferred to the solid. Again, in accordance with this mechanism, creep is expected to increase with temperature due to the effect of the latter on the viscosity of the water.

8.3.3. Movement of Interlayer Water

The movement of interlayer water, in and out of the laminated structure of the gel particles, was suggested to explain shrinkage and swelling of the cement paste (see section 7.3.4). Similarly, it has been suggested that creep is attributable to the same mechanism in which the exit of the interlayer water is brought about by the imposed external load, and not by the decrease in ambient humidity [8.5]. The exit of the interlayer water reduces the spacing between the layers, and thereby causes volume decrease, i.e. creep. On unloading, some of the water re-enters the structure, the spacing between the layers is increased and some of the creep is recovered. It should be pointed out, however, that in a later study it was concluded that this mechanism of water movement, although it occurs, is not the major mechanism involved [8.6].

8.3.4. Concluding Remarks

The three preceding mechanisms differ considerably, but all three attribute creep, in one way or another, to movement of water within the cement paste. In this respect, it may be noted that shrinkage is also attributable to movement of water. However, whereas in the case of creep, the movement of the water takes place within the paste, in the case of shrinkage the moisture exchange takes place between the paste and its surroundings.

Other mechanisms have been suggested to explain creep [8.7]. Nevertheless, it seems that the creep mechanism is not fully understood, and the suggested mechanisms do not always account for some of the creep aspects. For example, considering the preceding mechanisms, all three predict that no creep is to be expected in a saturated or in a completely dried paste. This is, however, not necessarily the case (see section 8.4.2.3).

8.4. FACTORS AFFECTING CREEP

8.4.1. Environmental Factors

It was pointed out earlier that the simultaneous drying of concrete increases creep, and that this increase is referred to as drying creep. Hence, it is to be expected that all factors which affect drying and induce shrinkage will similarly affect creep. It is further to be expected that creep will increase with the intensity of drying conditions, i.e. with the decrease in ambient humidity and the increase in temperature and wind velocity.

The effect of simultaneous drying (i.e. shrinkage) on creep is demonstrated in Fig. 8.3, and it is clearly evident that a more intensive drying (i.e. lower ambient relative humidity) brings about greater creep. This effect has been confirmed in many tests and is reflected, for example, in estimating creep with respect to ambient relative humidity in accordance with British Standard BS 8110, Part 2, 1985 (Fig. 8.4). Furthermore, it was suggested that, accordingly, the relation between total creep, C, and simultaneous shrinkage, S_s, may be

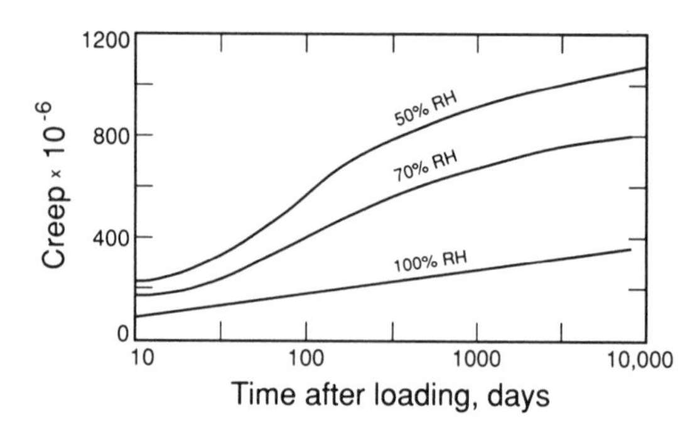

Fig. 8.3. Effect of simultaneous drying on creep of concrete moist cured for 28 days and then loaded and exposed to the relative humidities indicated. (Adapted from Ref. 8.8.)

expressed by the following expression [8.9]:

$$C = C_b (1 + kS_s) \tag{8.1}$$

in which C_b is the basic creep, S_s is the simultaneous shrinkage at the conditions considered and k is a constant which depends on concrete properties.

Considering that temperature affects the rate of drying, and thereby shrinkage, it is to be expected that creep also will increase with the rise in temperature. Moreover, noting that creep is associated with water movement within the cement, and that the viscosity of the water decreases with temperature, it is to be expected, again, that creep will increase with the rise in temperature.

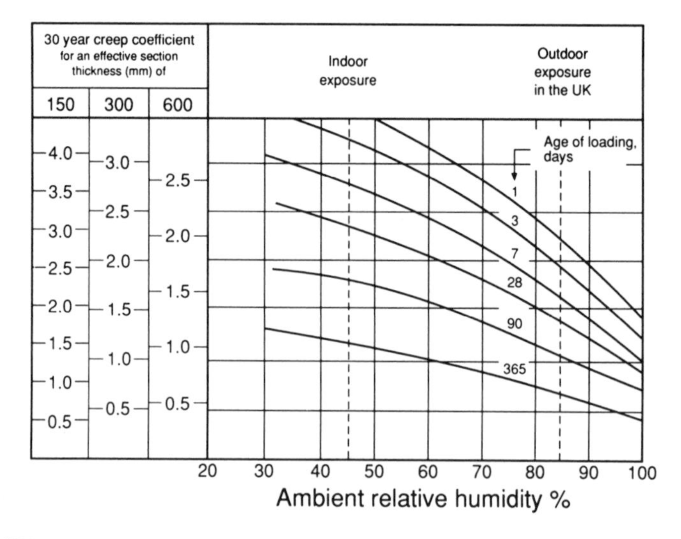

Fig. 8.4. Effects of relative humidity, age of loading and section thickness upon the creep factor. (Adapted from BS 8110, Part 2, 1985.)

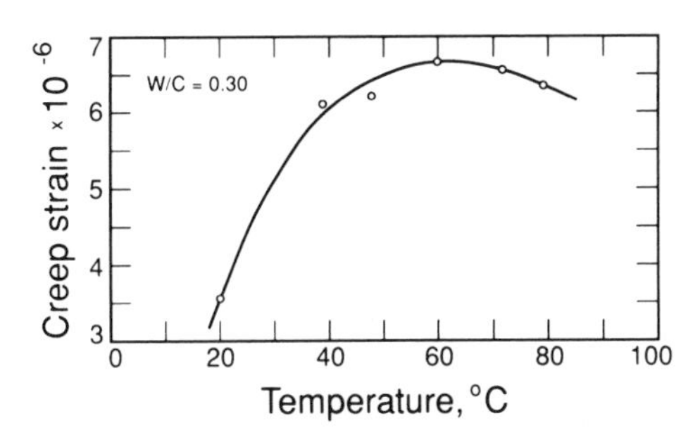

Fig. 8.5. Effect of ambient temperature on basic creep of cement paste loaded for 6 days at the age of 28 days. Applied stress 0·1 MP_a. (Adapted from Ref. 8.10.)

It can be seen from Fig. 8.5 that, indeed, creep increases with temperature. This increase, however, takes place up to the temperature of, say 60°C, but a further increase in temperature brings about a reversed trend. Such a reversed trend, at approximately 70°C, has been observed by others [8.11], and can be attributed to the two opposing effects of temperature. As already pointed out, the decreased viscosity of water is expected to increase creep. On the other hand, as will be seen later (see section 8.4.2.2), creep is strength related and, under otherwise the same conditions, a lower creep is to be expected in a stronger concrete. That is, as the rise in temperature accelerates hydration and thereby strength development, creep is expected to decrease with temperature. Apparently, the effect of the increased strength on creep, in the lower temperature range, is less than the effect of the decreased water viscosity. Hence, the increase in creep in the lower temperature range. In the higher range, however, the net effect of the two opposing effects is reversed, and creep decreases with a rise in temperature. It must be realised that in hot environments this reversed trend is of no practical importance because temperatures exceeding 60–70°C do not occur even under severe climatic conditions. Hence, even under such conditions, temperature may be considered to increase creep.

It was shown above that early and short exposure of fresh concrete to intensive drying increases strength (Chapter 6, Fig. 6.15) and reduces shrinkage (Chapter 7, Fig. 7.7). As both strength and shrinkage affect creep, it is to be expected that the same exposure will similarly affect creep, i.e. creep will be reduced when similarly exposed. This expected behaviour is confirmed by the data presented in Fig. 8.6 and supported by the data of some others [8.14]. It must be stressed again, however, that this apparent beneficial effect should not be considered as a possible recommendation to expose fresh concrete to early and intensive drying. From reasons elaborated earlier, such an exposure must definitely be avoided and the fresh concrete must be protected from drying as early as possible.

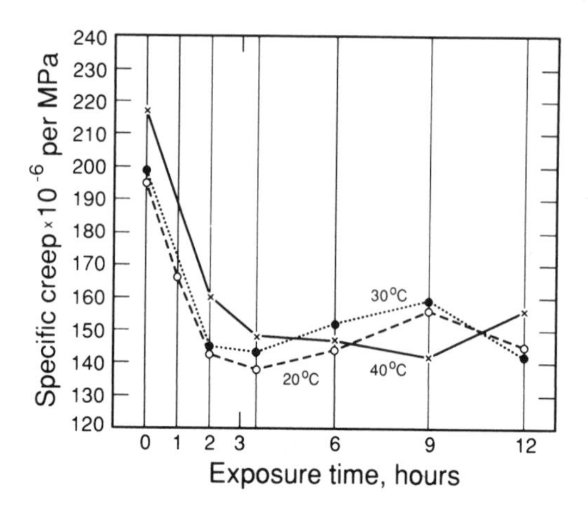

Fig. 8.6. Effect of early exposure, at the temperatures and relative humidities indicated (wind velocity 20 km/h), on specific creep of concrete at the age of 425 days. Concrete containing 350 kg/m^3 ordinary Portland cement (OPC) loaded at the age of 60 days and kept at 20°C and 65% RH. (Adapted from Refs 8.12 and 8.13.)

8.4.2. Concrete Composition and Properties

8.4.2.1. Aggregate Concentration and Rigidity

The aggregates normally used in concrete production do not creep, and the creep of concrete is determined, therefore, by the creep of the cement paste and its relative content in the concrete. It follows that a higher creep is to be expected in cement-rich concrete or, alternatively, creep is expected to increase with the decrease in aggregate concentration. This latter conclusion is confirmed by the data of Fig. 8.7.

As normal aggregates do not creep, their presence in the concrete restrains the creep of the paste to an extent which depends on their rigidity. Hence, for otherwise the same conditions, concretes made of soft aggregates are expected

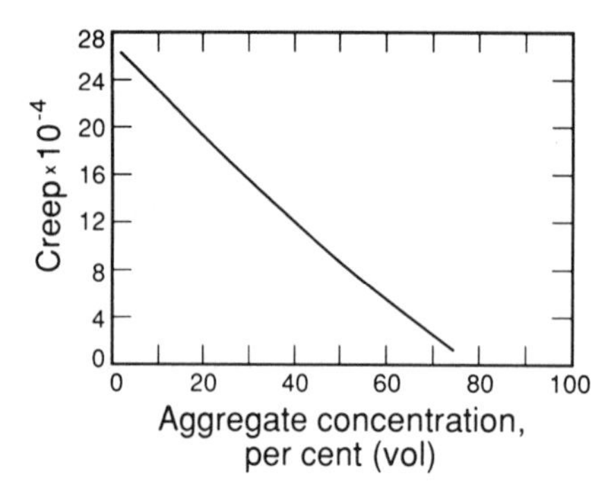

Fig. 8.7. Effect of aggregate concentration on creep of concrete loaded for 60 days at the age of 14 days. (Adapted from Ref. 8.15.)

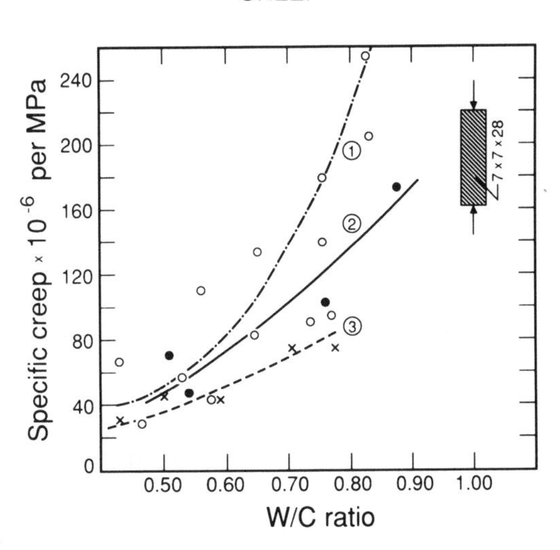

Fig. 8.8. Creep of concretes of different W/C ratios made with lightweight and normal-weight aggregates. (1) Air-entrained lightweight aggregate concrete, (2) as (1) but with no air entrainment, (3) normal-weight concrete. (Adapted from Ref. 8.16.)

to exhibit higher creep than those made with hard aggregates. Lightweight aggregate is softer than normal-weight aggregate. Hence, it follows that creep of lightweight aggregate concrete will be higher than that of normal weight aggregate concrete. This conclusion is confirmed by the data of Fig. 8.8.

The data of Fig. 8.8 compare creep of concretes made with the same water to cement (W/C) ratio. On the other hand, when concretes of the same strength are compared, essentially the same creep is observed (Fig. 8.9). The strength of lightweight aggregate concrete is lower than the strength of

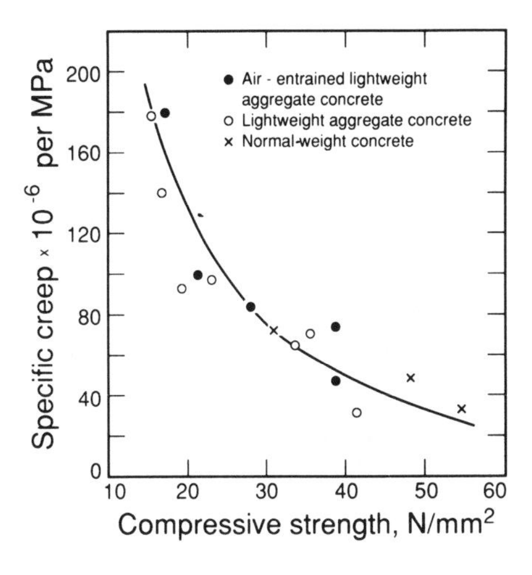

Fig. 8.9. Creep of concretes of different strengths made with lightweight and normal-weight aggregates. (Adapted from Ref. 8.16.)

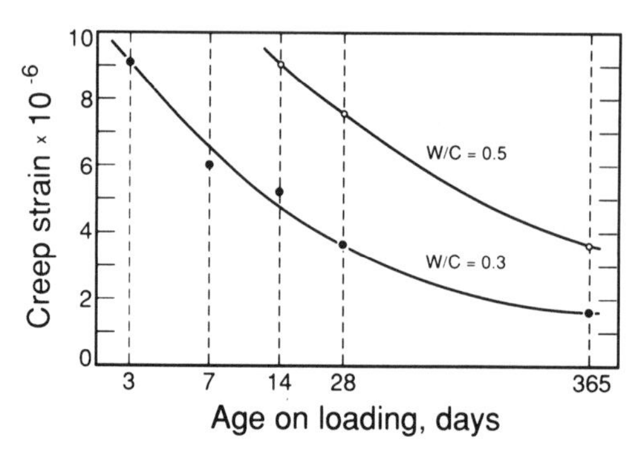

Fig. 8.10. Effect of W/C ratio on basic creep of cement paste after 6 days of loading. Applied stress 0·1 MP_a. (Adapted from Ref. 8.10.)

normal-weight concrete of the same W/C ratio (Chapter 6, Fig. 6.6) and, in order to obtain the same strength, the former concrete must be prepared with a lower W/C ratio than the latter one. The lower W/C ratio reduces the creep of the cement paste (see section 8.4.2.2), and this reduction counteracts the increased creep which is brought about by the use of the softer lightweight aggregate. Hence, essentially the same creep is exhibited by lightweight and normal-weight aggregate concretes of the same strength.

In view of the preceding discussion, it is evident that the effect of aggregate concentration and rigidity on creep must be similar to their effect on shrinkage. Indeed, creep of concrete can be expressed by the following equation, which is analogous to the one expressing shrinkage (see eqn (7.1)):

$$C = C_p(1 - V_a)^n \tag{8.2}$$

in which C and C_p are the creep of concrete and paste, respectively; V_a is the volume fraction of the aggregate, and n is a factor which depends on the elastic properties of the aggregate.

8.4.2.2. Strength, Stress and Stress to Strength Ratio

It is implied by the suggested creep mechanisms (see section 8.3), that creep must decrease with the increase in concrete modulus of elasticity and the increase in the stress level induced by the external load. The effect of modulus of elasticity and that of the stress level are self-evident once creep is considered as a delayed elastic deformation (see section 8.3.2). The modulus of elasticity is strength related, whereas strength is determined by the W/C ratio. Accordingly, Figs 8.8 and 8.10 indicate that, indeed, creep depends on the W/C ratio

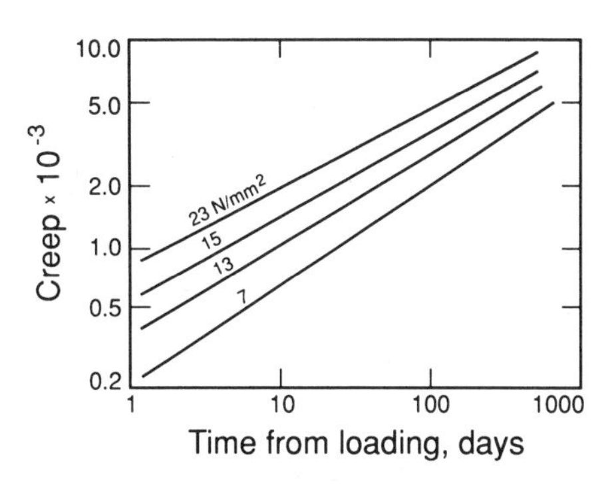

Fig. 8.11. Effect of stress level on creep of cement paste. (Adapted from Ref. 8.7.)

or, alternatively, on strength (Fig. 8.9). Similarly, the expected effect of the stress level is demonstrated in Fig. 8.11.

Noting that creep increases with the stress level and decreases with strength, it is to be expected that creep will increase with an increase in the stress to strength ratio. The data of Fig. 8.12 confirm this conclusion, and indicate that a linear relation between creep and stress to strength ratio exists up to the ratio of 0·85. Other values have been reported but, generally, this relation may be assumed to be linear up to the ratio of 0·3–0·4.

8.4.2.3. Moisture Content

The effect of moisture content on basic creep of cement paste is demonstrated in Fig. 8.13. The pastes in question (W/C = 0·4) were loaded (stress to strength ratio = 0·2) after reaching equilibrium with the surrounding atmosphere at the relative humidities indicated. It is evident from this figure that creep increases with an increase in ambient relative humidity, i.e. with an increase in moisture content of the paste. It may be pointed out that this observation is not in full agreement with the creep mechanisms described earlier (see section 8.3) which attribute creep to movement of water within the paste. As no such movement can take place in the absence of water, or when the paste is completely saturated, no creep is to be expected under such

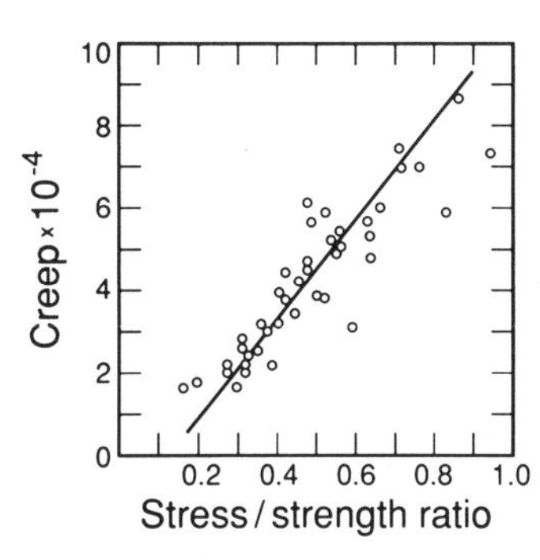

Fig. 8.12. Effect of stress to strength ratio on basic creep of cement mortars. (Adapted from Ref. 8.17.)

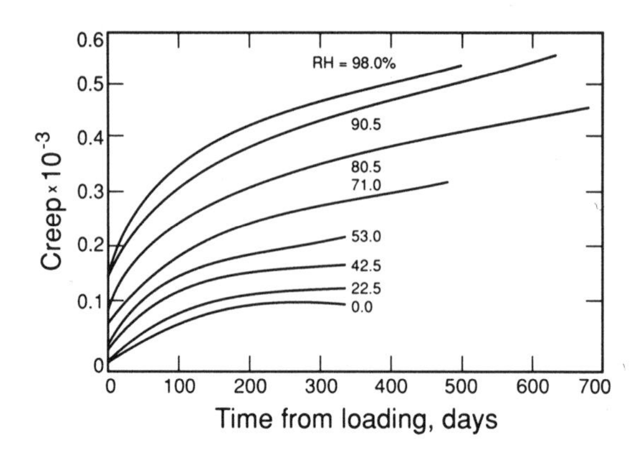

Fig. 8.13. Effect of ambient humidity on basic creep of cement paste. (Adapted from Ref. 8.18.)

conditions. This is not indicated by the data of Fig. 8.13 implying, in turn, that some other mechanisms may be involved.

8.4.2.4. Mineral Admixtures

In discussing the effect of mineral admixtures on drying shrinkage (see section 7.4.2.6), it was pointed out that such admixtures are expected to increase drying shrinkage because their presence gives the cement paste a finer pore structure. This expected behaviour, however, is not always supported by the available experimental data, and the sometimes contradictory nature of the test results involved was attributed to differences in testing conditions. Nevertheless, noting that shrinkage and creep mechanisms are both of a similar nature, mineral admixtures are expected to increase creep as well. In the case of creep, as indicated in Figs 8.14 and 8.15, test data support the expected effect, at least when fly-ash and granulated blast-furnace slag are considered.

8.5. SUMMARY AND CONCLUDING REMARKS

Creep is time-dependent deformation due to sustained loading. 'Basic creep' is the creep occurring in concrete at hygral equilibrium with ambient relative humidity. Simultaneous drying (i.e. shrinkage) increases creep, and the difference between the latter and basic creep is known as 'drying creep'. In practice, however, no such distinction is made and the term 'creep' is used indiscrimi-

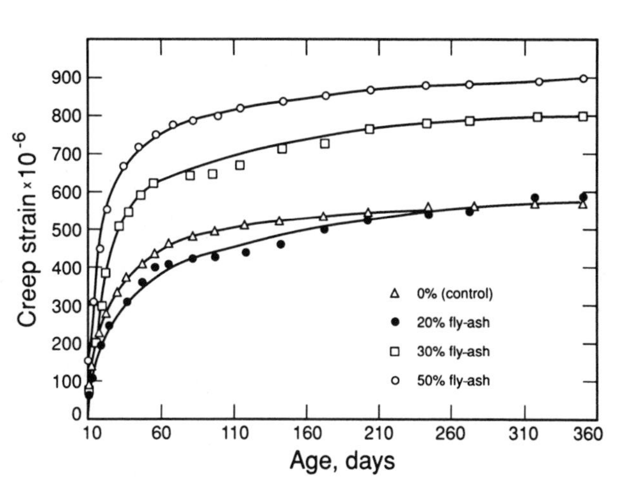

Fig. 8.14. Effect of replacing OPC with high-calcium fly-ash on creep of concrete. (Adapted from Ref. 8.19.)

nately whether or not drying is taking place. Creep is partly irrecoverable. Hence, the distinction between 'reversible' and 'irreversible' creep.

A few mechanisms have been suggested to explain creep, and most of them attribute creep to movement of water inside the cement paste. Creep increases with the increase in the intensity of drying conditions, i.e. with an increase in temperature and wind velocity and a decrease in relative humidity. Creep is also increased with stress to strength ratio and with an increase in moisture content. High-calcium fly-ash and granulated blast-furnace slag tend to increase creep. On the other hand, creep decreases with an increase in aggregate concentration and rigidity.

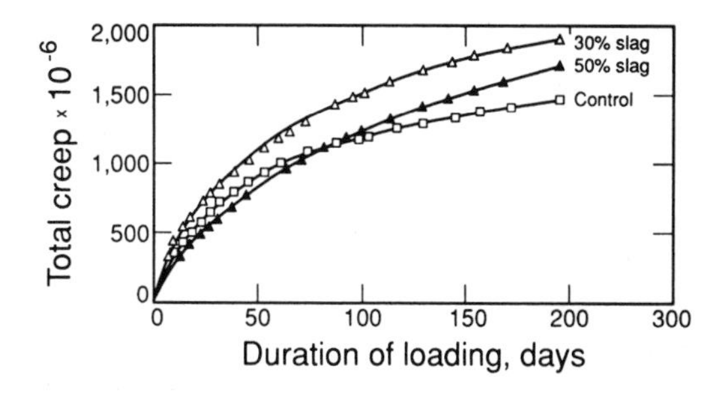

Fig. 8.15. Effect of replacing OPC with granulated blast-furnace slag on creep of concrete. (Adapted from Ref. 8.20.)

REFERENCES

8.1. Powers, T.C., Mechanism of shrinkage and reversible creep of hardened cement paste. In *Proc. Conf. Structure of Concrete and Its Behaviour Under Load*. London, 1965, Cement and Concrete Association, London, UK, 1968, pp. 319–44.
8.2. Bazant, Z.P., Delayed thermal dilation of cement paste and concrete due to mass transport. *Nuclear Engng Design*, **14** (1970), 308–18.
8.3. Ishai, O., Time-dependent deformational behaviour of cement paste, mortar and concrete. In *Proc. Conf. Structure of Concrete and Its Behaviour Under Load*. London, 1965, Cement and Concrete Association, London, UK, 1968, pp. 345–64.
8.4. Glucklich, J. & Ishai, O., Creep mechanism in cement mortar. *Proc. ACI*, **59**(7) (1962), 923–48.
8.5. Feldman, R.F. & Sereda, J.P., A new model for hydrated cement and its practical implications. *Engng J.*, **53** (1970), 53–9.
8.6. Feldman, R.F., Mechanism of creep of hydrated Portland cement. *Cement Concrete Res.*, **17**(50) (1972), 521–40.
8.7. Wittmann, F. H., Discussion of some factors influencing creep of concrete. Research Series III – Building, No. 167, The State Institute for Technical Research, Finland, 1971.
8.8. Troxell, G.E., Raphael, J.M. & Davis, R.E., Long-time creep and shrinkage tests of plain and reinforced concrete. *Proc. ASTM*, **58** (1958), 1101–20.
8.9. L'Hermite, R., Current ideas about concrete technology. Documentation Technique du Batiment et des Travaux Publics, Paris, France, 1955 (in French).
8.10. Ruetz, W., A hypothesis for creep of hardened cement paste and the influence of simultaneous shrinkage. In *Proc. Conf. Structure of Concrete and Its Behaviour Under Load*. Cement and Concrete Association, London, UK, 1968, pp. 365–403.
8.11. Neville, A.M., *Properties of Concrete* (3rd edn). Longman Scientific & Technical, UK, 1986, p. 411.
8.12. Jaegermann, C.H., Effect of exposure to high evaporation and elevated temperatures of fresh concrete on the shrinkage and creep characteristics of hardened concrete. DSc Thesis, Faculty of Civil Engineering, Technion – Israel Institute of Technology, Haifa, Israel, July 1967 (in Hebrew with an English synopsis).
8.13. Jaegermann, C.H. & Glucklich, J., Effect of high evaporation during and shortly after casting on the creep behaviour of hardened concrete. *Mater. Struct.*, **2**(7) (1967), 59–70.
8.14. Shalon, R. & Berhane, Z., Shrinkage and creep of mortar and concrete as affected by hot-humid environment. In *Proc. RILEM Symp. on Concrete and*

Reinforced Concrete in Hot Countries, Haifa 1971, Vol. I, Building Research Station, Technion – Israel Institute of Technology, Haifa, pp. 309–21.

8.15. Neville, A.M., Creep of concrete as a function of the cement paste. *Mag. Concrete Res.*, **16**(46) (1964), 21–30.

8.16. Soroka, I. & Jaegermann, C.H., Properties and possible uses of concrete made with natural lightweight aggregate (Part One). Report to the Ministry of Housing, Building Research Station, Technion – Israel Institute of Technology, Haifa, Israel, 1972 (in Hebrew).

8.17. Neville, A.M., Tests on the influence of the properties of cement on the creep of mortars. *RILEM Bull.*, **4** (1959), 5–17.

8.18. Wittmann, F.H., The effect of moisture content on creep of hardened cement pastes. *Rheal. Acta*, **9**(2) (1970), 282–87 (in German).

8.19. Yuan, R.L. & Cook, J.E., Study of a class C fly ash concrete. In *Fly Ash, Silica Fume, Slag and Other Mineral By Products* (ACI Spec. Publ. SP 79, Vol. I), ed. V.M. Malhotra. ACI, Detroit, MI, USA, 1983, pp. 307–19.

8.20. Neville, A.M. & Brooks, J.J., Time dependent behaviour of cemsave concrete. *Concrete*, **9**(3) (1975) 36–9.

Chapter 9
Durability of Concrete

9.1. INTRODUCTION

The ability of concrete to withstand the damaging effects of environmental factors, and to perform satisfactorily under service conditions, is referred to as 'durability'. Clearly the durability of concrete is of prime importance in all engineering applications, and the satisfactory performance of the concrete must be ensured throughout its expected service life. Giving the concrete the required durability in aggressive environments is by no means easily achieved, and requires careful attention to details during all stages of its mix design and production. This is particularly the case under hot-weather conditions where environmental factors may further aggravate the problem, and make it more difficult for the concrete to attain the required quality.

Chemical corrosion of concrete, and that of the reinforcing steel as well, are conditional on the presence of water (moisture), and their intensity is very much dependent on concrete permeability. Dense and impermeable concrete reduces considerably the ingress of aggressive agents into the concrete, and thereby limits their corrosive attack to the surface only. The same applies to the penetration of air (i.e. oxygen and carbon dioxide) and chloride ions, both which play an important role in the corrosion of the reinforcing steel. Porous concrete, on the other hand, allows the aggressive water to penetrate, and the attack proceeds simultaneously throughout the whole mass. Hence, such an attack is much more severe. Similarly, a porous concrete allows air and chloride ions to reach the level of the reinforcement, and thereby promotes corrosion in the steel bars. Hence, durability-wise, and regardless of the specific conditions involved, dense and impermeable concrete is always required when the latter is intended for use in aggressive environments. In view of its general relevance, the discussion of permeability precedes that of the corrosion of the concrete and the reinforcing steel.

Finally, concrete deterioration may be caused by different aggressive agents and processes. The following discussion is of a limited nature and includes only the more important ones which are also relevant to hot weather conditions. A more detailed discussion can be found elsewhere [9.1,9.2].

9.2. PERMEABILITY

9.2.1. Effect of Water to Cement (W/C) Ratio

The porosity of concrete aggregates usually does not exceed 1–2%, whereas that of hardened cement is very much greater and, depending on the W/C ratio and the degree of hydration, is of the order of some 50% [9.3]. Consequently, the permeability of concrete is determined by the permeability of the set cement which, in turn, is determined by its porosity or rather by the continuous part of its pore system. The very small gel pores do not allow the passage of water and, consequently, permeability is conditional on the presence of bigger pores, namely, the capillary pores. Capillary porosity, in turn, is determined by the W/C ratio and the degree of hydration. Hence, for the same degree of hydration (i.e. the same age and curing regime) permeability is determined by the W/C ratio alone.

The relation between the W/C ratio and permeability is described in Fig. 9.1. It may be noted that for W/C ratios below, say 0·45, permeability is rather low and is hardly affected by further reductions in the W/C ratio. At higher ratios, however, permeability becomes highly dependent on the W/C ratio, and a comparatively small increase in the latter is associated with a considerable increase in the former. This change in the relationship is attributable to a change in the nature of the pore system. In the lower W/C ratio range, the system is discontinuous and the capillary pores are separated from each other by the cement gel. The permeability of the gel being rather low, the permeability of the concrete as a whole is similarly low and independent of capillary porosity. In the higher W/C ratio range, the pore system is continuous and allows, therefore, the passage of water. Hence, increasing the pore

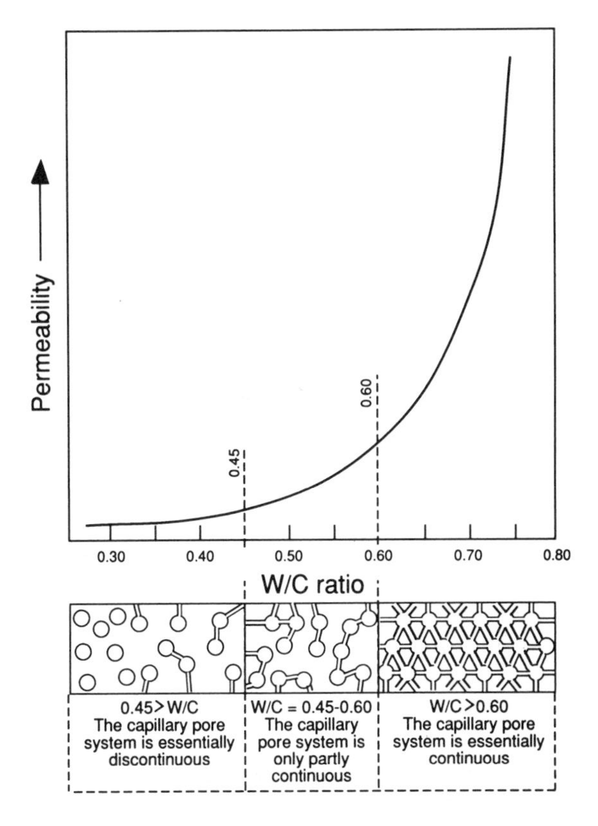

Fig. 9.1. The effect of W/C ratio on nature of pore structure and permeability of concrete.

volume in such a system increases permeability. As the porosity is determined by the W/C ratio, permeability is increased with an increase in the W/C ratio.

It may be concluded from Fig 9.1 that a W/C ratio of 0·45 or less produces virtually impermeable concrete. Indeed, this conclusion is applied in everyday practice when a dense and durable concrete is required, and is reflected, for example, in ACI recommendations (Tables 9.1 and 9.2). This conclusion, however, is valid only for well-cured concrete because even with a relatively low W/C ratio, concrete may have a continuous pore system if the cement is not sufficiently hydrated. In this context, the importance of adequate curing cannot be over-emphasised.

Table 9.1. Maximum Permissible W/C or Water/Cementitious Materials[a] Ratios for Concrete in Severe Exposures.[b]

Type of structure	*Structure continuously or frequently wet and exposed to freezing and thawing*[c]	*Structure exposed to sea water or sulphates*
Thin sections, (railings, curbs, sill, ledges, ornamental work) and sections with less than 25 mm cover over the steel	0·45	0·40[d]
All other structures	0·50	0·45[d]

[a] *Materials should conform to ASTM C618 and C989.*
[b] *Adapted from Ref. 9.4.*
[c] *Concrete should also be air entrained.*
[d] *If sulphate-resisting (types II or V of ASTM C150) is used, permissible W/C or water/cementitious materials ratio may be increased by 0·05.*

Table 9.2. Recommendations for Sulphate-Resistant Normal-Weight Concrete.[a]

Intensity of sulphate attack	*Sulphate concentration,* SO_4^{2-}		*Maximum W/C ratio*[b]	*Type of cement*[c]
	In soil, water soluble (%)	*In water (ppm)*		
Mild[d]	0·00–0·10	0–150	—	—
Moderate[e]	0·10–0·20	150–1500	0·50	Type II or portland-pozzolan or portland slag
Severe	0·20–2·00	1500–10 000	0·45	Type V
Very severe	Over 2·00	Over 10 000	0.45	Type V with a pozzolana[f]

[a] *Adapted from Ref. 9.5.*
[b] *A lower W/C ratio may be necessary to prevent corrosion of the reinforcement (see Table 9.1).*
[c] *Designation in accordance with ASTM C150 (section 1.5).*
[d] *Negligible attack: no protective means are required.*
[e] *Seawater also falls in this category (see following discussion).*
[f] *Only a pozzolan which has been determined by tests to improve sulphate resistance when used in concrete containing type V cement (see following discussion).*

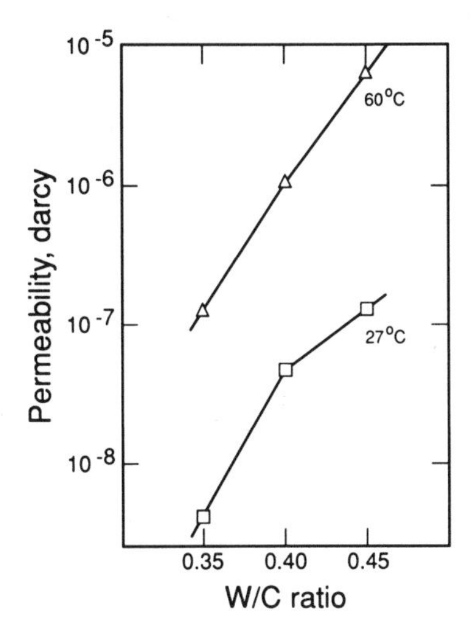

Fig. 9.2. Effect of temperature and W/C ratio on permeability of cement paste at the age of 28 days. (Adapted from Ref. 9.6.)

9.2.2. Effect of Temperature

It was demonstrated earlier (see section 2.5.4) that temperature affects pore-size distribution, and exposing the hydrating cement to higher temperatures brings about a coarser pore system. As permeability is mainly determined by the coarser pores (i.e. capillary pores), it is to be expected that, under otherwise the same conditions, permeability will increase with temperature. This is confirmed by the experimental data presented in Figs 9.2 and 9.3 implying that, under hot-weather conditions, a concrete of greater permeability, and

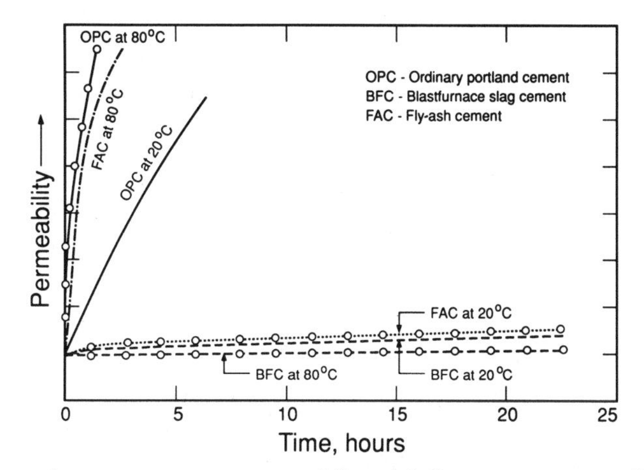

Fig. 9.3. Effect of temperature on permeability of 1:2 cement mortars (W/C = 0·65) made with different types of cement. (Adapted from Ref. 9.7.)

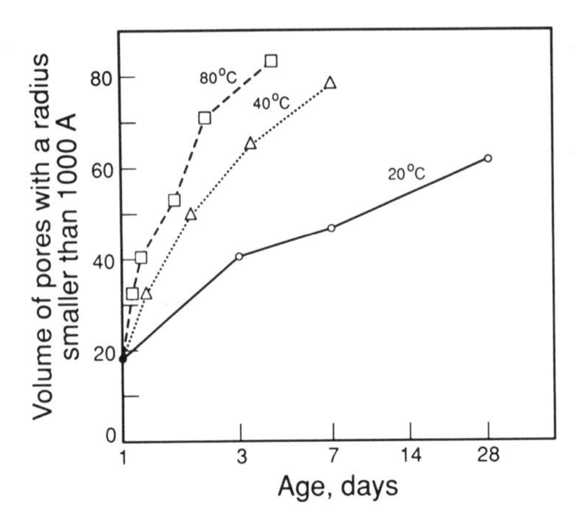

Fig. 9.4. Effect of temperature on volume percentage of pores having a radius smaller than 1000Å in ISO mortars made of blended cement containing 62·5% slag. (Adapted from Ref. 9.11.)

therefore, of a greater sensitivity to attack by aggressive agents, is to be expected.

Mineral admixtures, such as blast-furnace slag, silica fume and fly-ash, were shown to produce concrete of a finer pore structure and a lower permeability, although not necessarily with a lower porosity [9.8–9.10]. This reduced permeability brought about by the use of admixtures is demonstrated, for example, in Fig. 9.3 which compares the permeability of ordinary Portland cement (OPC) mortar with the permeabilities of corresponding mortars made of slag and fly-ash cements. It can be seen that at 20°C the permeability of the mortars made with both blended cements tested was negligible, whereas that of the Portland cement mortar was rather high. Moreover, the permeability of the latter increased considerably when the mortar was hydrated at 80°C. In this respect it is of interest to note that the permeability of the mortar made with the fly-ash cement was similarly adversely affected. That is, the use of fly-ash cement, although very beneficial at 20°C, is not necessarily advantageous when permeability at elevated temperatures is considered. On the other hand, the permeability of the slag cement mortar was not affected by the elevated temperature of 80°C. Moreover, it was shown that, contrary to the effect of temperature on the porosity of Portland cement (Chapter 2, Fig. 2.12), the porosity of slag cement becomes finer with temperature (Fig. 9.4). Accordingly, when low permeability is required, the use of slag cement is to be preferred, and particularly under hot-weather conditions. It will be seen later that the use of slag cement may be desirable also for additional reasons. Indeed, such a cement, containing 65% slag, is sometimes recommended for use in hot regions [9.12].

9.2.3. Summary and Concluding Remarks

Permeability determines to an appreciable extent concrete durability and, consequently, a dense and impermeable concrete must be produced when a durable concrete is required, i.e. when the concrete is to be exposed to an aggressive environment. In turn, permeability is determined by the porosity of the cement paste, or rather by the continuous part of its capillary pore system. In a well-cured (hydrated) concrete, the latter becomes essentially discontinuous at the W/C ratio of, say, 0·45. Hence, such a W/C ratio is recommended for concrete in severe exposures (Tables 9.1 and 9.2).

Elevated temperatures, through their effect on pore-size distribution, increase permeability. In this respect, a blended cement containing 65% slag is preferable because the permeability of such a cement is not adversely affected by temperature. Moreover, the permeability of this cement at normal temperatures is lower, in the first instance, than that of OPC. Hence, the use of slag cement is sometimes recommended for use in hot environments.

9.3. SULPHATE ATTACK

Most sulphates are water-soluble and severely attack Portland cement concrete. A notable exception, in this respect, is barium sulphate (baryte) which is virtually insoluble in water and is, therefore, not aggressive with respect to concrete. In fact, barytes are used to produce heavy concrete which is sometimes used in the construction of atomic reactors and similar structures, because of its improved shielding properties against radioactive radiation.

The intensity of sulphate attack depends on many factors, such as the type of the sulphate involved, and its concentration in the aggressive water or soil, but under extreme conditions, it may cause severe damage, and even complete deterioration of the attacked concrete. In nature sulphates may be present in ground water and soils, and particularly in soils in arid zones. Sulphates are also present in seawater. The comparatively wide occurrence of sulphates, on the one hand, and the severe damage which sulphate attack may cause, on the other, makes this type of attack widespread and troublesome. Hence, it must be seriously considered in many engineering applications.

9.3.1. Mechanism

The mechanism of sulphate attack is not simple, and there still exists some controversy with respect to its exact nature. Generally, however, the sulphates react with the alumina-bearing phases of the hydrated cement to give a high-sulphate form of calcium aluminate ($3CaO.Al_2O_3.3CaSO_4.32H_2O$, i.e. $C_3A.3C\bar{S}.H_{32}$), known as ettringite.

The formation of ettringite due to sulphate attack, involves an increase in the volume of the reacting solids. Considering the porosity of the cement paste, it may be stipulated that this volume increase may take place without causing expansion. Indeed, this would have been the case if the reactions involved had occurred through solution, and the resulting products would have precipitated and crystallised in the available pores throughout the set cement. This, however, is not the case, and in practice sulphate attack of concrete is usually associated with expansion. It is generally accepted, therefore, that the reactions involved are of a topochemical nature (i.e. liquid–solid reactions) and occur on the surface of the aluminium-bearing phases. It is further argued that the space available locally where the reactions take place, is not great enough to accommodate the increase in the volume of the solids, and this volume constraint results in a pressure build-up. In turn, such a pressure causes expansion and, in the more severe cases, cracking and deterioration.

9.3.2. Factors Affecting Sulphate Resistance

9.3.2.1. Cement Composition

In discussing the mechanism of sulphate attack, it was explained that the vulnerability of the concrete to such an attack is attributable to the presence of the alumina-bearing phases in the set cement. The alumina-bearing phases are the hydration products of the C_3A of the cement. It follows that the sulphate resistance of the cement will increase with a decrease in its C_3A content. Indeed, this conclusion has been confirmed by both field and laboratory tests [9.13, 9.14], and constitutes the basis for the production of sulphate-resisting cement, i.e. Portland cement in which the C_3A content does not exceed 5% (cement type V in accordance with ASTM C150) (see section 1.5.3). The latter conclusion is demonstrated in Fig. 9.5 which presents the data of exposure tests which were carried out on concretes made with cements of different C_3A content. In Fig. 9.5 the intensity of the sulphate attack is expressed by the 'rate

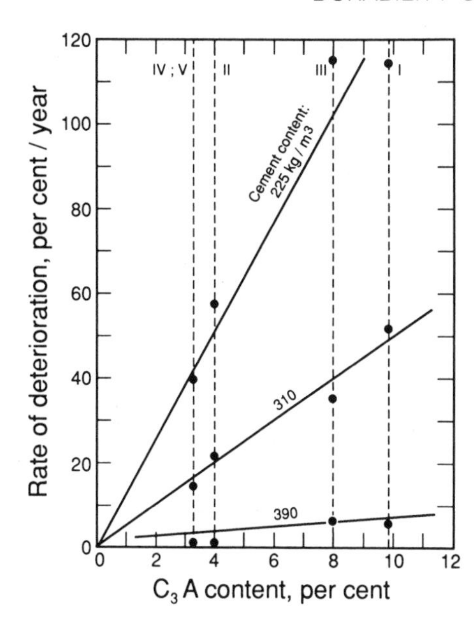

Fig. 9.5. Effect of the C_3A content in Portland cement on the rate of deterioration of concrete exposed to sulphate bearing soils. (Adapted from Ref. 9.14.)

of deterioration' (percent per year), and it is quite evident that this rate decreases with the decrease in the C_3A content of the cement.

9.3.2.2. Cement Content and W/C Ratio

In view of the improved resistance to sulphate attack, the use of sulphate-resisting cement is recommended when such an attack is to be considered, e.g. in concrete exposed to sulphate-bearing soils or sulphate-containing water (Table 9.2). On the other hand, it can be concluded from the very same data of Fig. 9.5, that the increased resistance to sulphate attack can be achieved by the use of a high cement content (i.e. a low W/C ratio) and not necessarily by the use of a low C_3A content cement. It can be seen, for example, that a cement content of 390 kg/m^3 imparts to the concrete a high sulphate resistance, apparently even higher than that which can be achieved by the use of a cement with a low C_3A content. In other words, in producing sulphate-resistant concrete, the use of sulphate-resisting cement must be combined with a specified minimum cement content. Indeed, this conclusion is reflected, for example, in BS 8110, Part 1, 1985, which specifies such a minimum. In accordance with conditions of exposure and maximum size of aggregate particles, this specified minimum varies between 280 and 380 kg/m^3.

The cement content affects the sulphate-resisting properties of concrete, mainly through its effect on the W/C ratio. That is, under otherwise the same conditions, an increase in the cement content reduces the W/C ratio. The

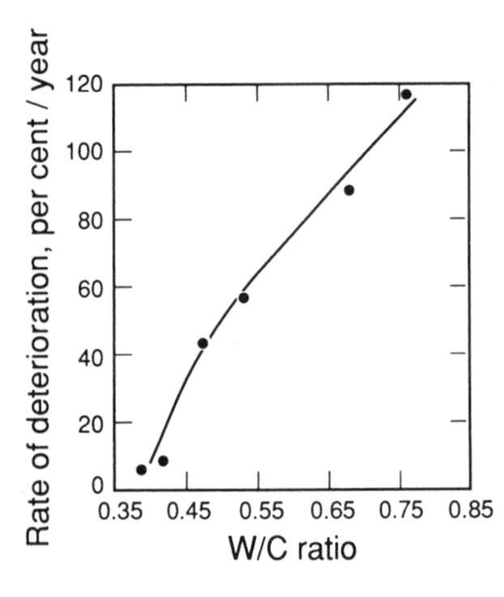

Fig. 9.6. Effect of W/C ratio on rate of deterioration of concrete made of ordinary Portland cement and exposed to sulphate bearing soils. (Adapted from Ref. 9.14.)

reduced W/C ratio, in turn, reduces concrete permeability, and thereby improves its sulphate-resisting properties. This effect of the W/C ratio is indicated by the data of Fig. 9.6, suggesting that in order to produce a sulphate-resistant concrete a W/C ratio of, say, 0·40, must be selected. Indeed, this ratio is recommended when OPC is used. If, however, a sulphate-resisting cement is used, a somewhat greater W/C ratio may be adopted, i.e. 0·45 (Table 9.2).

The reduction of the calcium hydroxide content in the set cement is important when the source of the sulphate ions is other than gypsum because the latter ions react, in the first instance, with the calcium hydroxide. This is usually the case when the SO_4^{2-} concentration in the aggressive water exceeds some 1500 mg/litre because the solubility of gypsum in water at normal temperatures is rather low, being approximately 1400 mg/litre. Calcium hydroxide is produced as a result of the hydration of both the Alite (C_3S) and the Belite (C_2S) of the cement. The hydration of the Alite, however, produces considerably more calcium hydroxide than the hydration of the Belite (see section 2.3). Hence, in this respect, a cement low in C_3S is to be preferred. It may be noted that, sometimes, sulphate-resisting cements are characterised by a low C_3S content (Chapter 1, Table 1.4).

9.3.2.3. Pozzolans

It was explained earlier (see section 3.1.2) that pozzolans react with lime in the presence of water at room temperature. Hence, the concentration of the calcium hydroxide in hydrated blends of Portland cement and a pozzolan is lower than in hydrated unblended cements. It is to be expected, therefore, that the use of Portland–pozzolan cement, or the addition of a pozzolan to the mix,

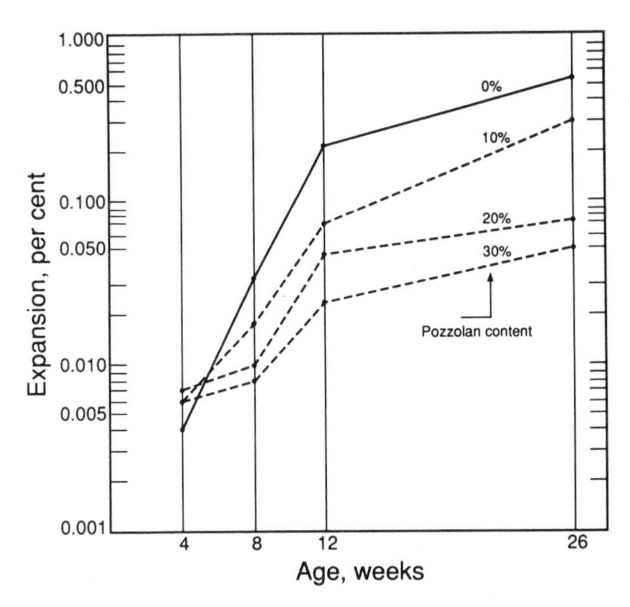

Fig. 9.7. Effect of Santorin earth on expansion of 1″ × 1″ × 10″ (25·4 mm × 25·4 mm × 254 mm) mortar prisms immersed in 10% Na_2SO_4 solution. (Adapted from Ref. 9.18.)

would produce concrete of improved sulphate-resisting properties. Moreover, such an improvement may also be expected in view of the finer pore system, and the lower permeability which are associated with the use of pozzolans. Yet another reason is the diluting effect of the partial replacement of Portland cement on the C_3A concentration. This expected beneficial effect of pozzolans on sulphate resistance of concrete is well recognised and has been confirmed by many studies [9.15–9.17]. It is demonstrated here, for example, in Fig. 9.7 for natural pozzolan (Santorin earth) and in Fig. 9.8 for low-calcium fly-ash, where the vulnerability to sulphate attack is measured by the expansion of the test specimens due to immersion in sulphate solution. It can be seen that, indeed, the use of Santorin earth and some fly-ashes was associated with a lower expansion, i.e. with improved sulphate-resistance properties.

In view of the preceding discussion, the use of Portland–pozzolan cements and pozzolanic admixtures is recommended for concrete in order to control sulphate attack (Table 9.2). This recommendation is particularly relevant to conditions where the attack of alkali sulphates is to be considered, and a lower concentration of calcium hydroxide is, therefore, desired. In this respect it must be pointed out that the preceding discussion and conclusions are not necessarily valid when sulphate-resisting cements are used. It will be explained below (see

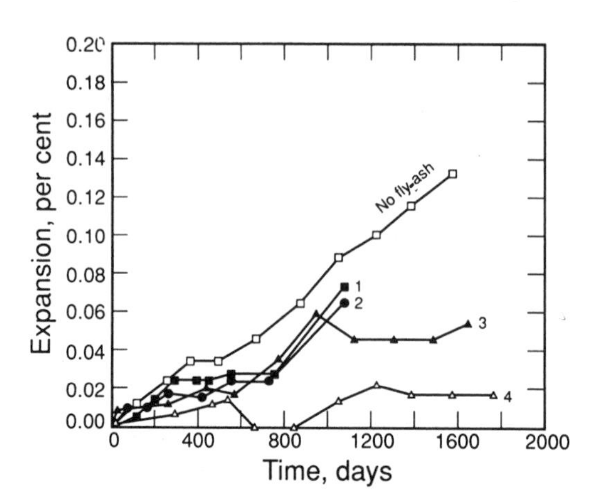

Fig. 9.8. Sulphate expansion of concrete containing low-calcium fly-ash of different compositions marked 1 to 4. (Adapted from Ref. 9.19.)

section 9.3.3) that for these types of cements only certain types of pozzolans may be useful.

9.3.2.4. Blast-Furnace Slag

Generally, replacing a substantial part of Portland cement with blast-furnace slag improves the sulphate-resisting properties of concrete. This effect is demonstrated, for example, in Fig. 9.9, and has been observed by others as well [9.21]. Granulated blast-furnace slag usually does not react with calcium hydroxide. Hence, the improvement in sulphate-resisting properties cannot be attributed to the reduced $Ca(OH)_2$ concentration due to the latter reaction, but rather to the diluting effect which is brought about by replacing a substantial part of Portland cement with slag. On the other hand, the concentration of the alumina-bearing phases is only partly affected by the latter replacement because calcium aluminates are produced in the hydration of the slag (see section 3.1.3.1). Hence, the improved sulphate properties of blended slag cements are mainly attributed to the finer pore system which characterises such cements (Chapter 3, Fig. 3.15). In Fig. 9.9 the effect of sulphate attack is measured by its effect on the flexural strength of the specimens tested. It can be seen that once the slag content exceeded some 65%, the immersion in the sulphate solution virtually did not affect strength, whereas at lower contents the specimens were actually destroyed, i.e. the relative strength equalled zero. Hence, slag cements with a slag content of 65–70% or more, are recommended for use in controlling sulphate attack.

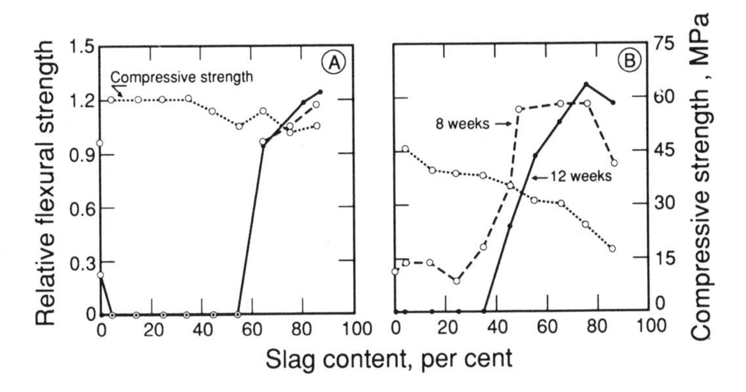

Fig. 9.9. Effect of slag content on flexural strength of 1:3 cement mortars immersed at the age of 21 days for 8 and 12 weeks in a 4·4% Na_2SO_4 solution. Relative flexural strength is expressed as the ratio of the strength of the mortars immersed in the sulphate solution to the corresponding strength of the mortars immersed in water. C_3A of the cement 11% and its fineness 300 m^2/kg. Alumina content of the slag(A) – 17·7%, and of slag (B) – 11·1%. Fineness of slags 500 m^2/kg. (Adapted from Ref. 9.20.)

9.3.2.5. Temperature

It was shown earlier (see section 2.5.1) that chemical reactions are considerably accelerated with temperature. Hence, it is to be expected that the intensity of sulphate attack would increase with temperature as well. In practice, however, the expansion of concrete due to sulphate attack, and its associated damaging effect, do not increase with temperature. In fact, as can be seen in Fig. 9.10, the opposite occurs and sulphate expansion actually decreases with the rise in temperature. This decrease is attributable to the nature of the chemical reactions which take place under elevated temperatures. Apparently, due to the

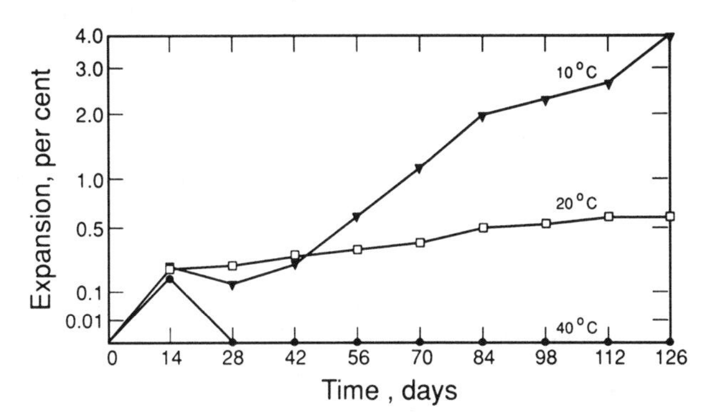

Fig. 9.10. Effect of temperature on the expansion of cement mortar exposed to sodium sulphate solution. (Adapted from Ref. 9.21.)

increased solubility of the sulphates and the ettringite, a greater part of the reactions occur through solution and less ettringite is deposited topochemically. Consequently, less pressure is generated due to the restrained volume increase, expansion is thereby reduced, and less damage occurs.

9.3.3. Controlling Sulphate Attack

In view of the preceding discussion, it may be concluded that in order to produce concrete sulphate-resisting properties, a suitable cement, combined with a low W/C ratio (or, alternatively, with a minimum cement content) should be used. These conclusions are summarised in Table 9.2 in accordance with American practice (ACI Committee 201), but similar recommendations are specified in many other codes (e.g. BS 8110, Part 1, 1985). It may be noted that the intensity of the sulphate attack is classified only with respect to the sulphate concentration in the aggressive water or in the soil, whereas the intensity of the attack is also determined by other factors such as type of the sulphate involved, and the nature of the contact between the concrete and the aggressive water, i.e. continuous immersion or alternate cycles of wetting and drying. It is rather difficult, however, to allow for all the factors involved, and that is why the classification of the intensity of sulphate attack is usually based solely on sulphate concentration.

The salt content of seawater usually varies between 3·6 and 4·0% of which some 10% are sulphates, namely magnesium sulphate ($MgSO_4$), gypsum ($CaSO_4$) and potassium sulphate (K_2SO_4). Accordingly, the sulphate concentration in seawater may reach 4·0 mg/litre which is equivalent to a SO_4^{2-} concentration greater than 2500 mg/litre. Hence, in accordance with Table 9.2, a 'severe' sulphate attack is to be expected. Nevertheless, experience has shown that the corrosion of concrete in seawater is much smaller than would be expected from its sulphate concentration explaining, in turn, why the attack of seawater is considered to be only 'moderate' in Table 9.2 (see footnote *e*). The exact reason for the reduced aggressiveness of sulphates in seawater is not completely clear. It has been suggested, for example, that the greater solubility of ettringite and gypsum in chloride solutions reduces the effect of the volume increase which is associated with sulphate attack [9.22]. Some other explanations have been offered [9.23, 9.24] but, regardless of the exact reason involved, sulphate attack of concrete exposed to seawater may be considered 'moderate' and treated accordingly.

It was mentioned earlier (see section 9.3.2.3), that only certain pozzolans improve the sulphate-resisting properties of concrete made from sulphate-resisting (type V) cements. Hence, the use of pozzolanic additions, which is recommended in Table 9.2 for 'very severe' exposure, is conditional on proving that, indeed, the pozzolan in question improves sulphate resistance of concrete when made of type V cement. Apparently, the effect of pozzolans on the latter property is related to their $SiO_2/(Al_2O_3 + Fe_2O_3)$ ratio (i.e. the ratio of the silica content to the combined contents of the alumina and the ferric oxide), and sulphate-resisting properties are improved only when pozzolans with a high ratio are used [9.25].

9.4. ALKALI–AGGREGATE REACTION

Normal aggregates are expected to be inert in the water–cement system, and this is usually the case. Some aggregates, however, may contain reactive components which, in the presence of water, may react with the alkalies of the cement (see section 1.3.4), or with alkalies from external sources. Consequently, expansion occurs which, under severe conditions, may cause the concrete severe damage and deterioration. The more common alkali–aggregate reaction involves reactive silicious materials and, accordingly, is referred to as 'alkali–silica reaction'. A much less common reaction involves carbonates and may occur with argillaceous (i.e. clay-containing) dolomitic limestones. Similarly, this reaction is referred to as 'alkali–carbonate reaction'. In this case a so-called 'dedolomitisation' process takes place (i.e., a process which is, essentially, the breaking down of the dolomite into calcium carbonate and magnesium hydroxide), and in the presence of clay this process may cause cracking and deterioration. The alkali–carbonate reaction has been observed to a very limited extent, and the following discussion, unless explicitly stated, relates, therefore, to the alkali–silica reaction alone. As a result of this latter reaction, an alkali–silica gel of the swelling type is formed which, on absorption of water, has its volume increased. Due to volume restraint within the concrete, pressure is generated which, in turn, may cause cracking and deterioration. Sometimes such cracking is accompanied by the exudation of the alkali–silica gel from the cracks, or by pop-outs and spalling on the surface of the effected concrete.

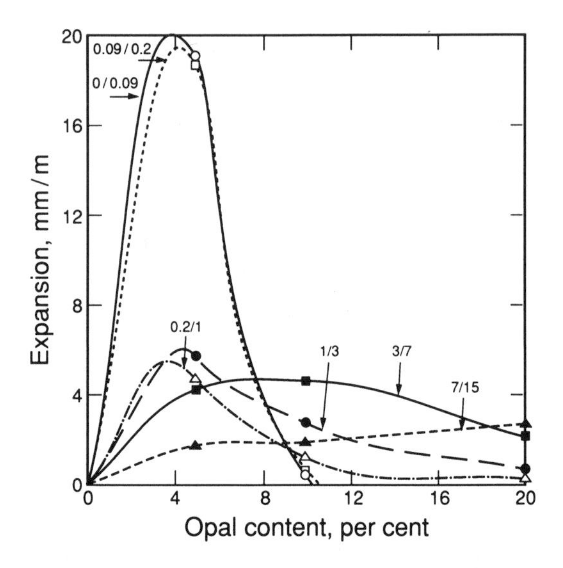

Fig. 9.11. The effect of opal content and particle size of the aggregate on the expansion of concrete due to alkali–silica reaction (particle size in mm). (Adapted from Ref. 9.26.)

9.4.1. Reactive Aggregates

It was pointed out earlier that the alkali–silica reaction involves the presence of reactive siliceous constituents in the aggregate, and such constituents may occur in opaline, siliceous limestones and many other rocks. (A list of potentially reactive aggregates and minerals can be found in Ref. 9.5.) In this respect, it must be realised that the presence of the minerals in question does not, necessarily, bring about alkali–silica reactions to the extent which may damage the concrete. This possible behaviour is due to the fact that the intensity of the alkali–silica reactions depends not only on the nature of the specific mineral involved but also, for example, on its concentration in the aggregate and its particle size (Fig. 9.11). Moreover, this dependence is not simple, and is usually characterised by a 'pessimum' content, i.e. a content which imparts the concrete maximum expansion. This is demonstrated in Fig. 9.11 in which the pessimum content of the opal considered is 4% when its particle size is less than 3 mm. Hence, the assessment of aggregate reactivity from its mineral and chemical composition reflects on its potential reactivity rather than on its actual performance in concrete. Further assessment can be based on additional tests, such as the one described in ASTM C227 and, of course, on past experience with the aggregates in question, or with aggregates of a similar origin and nature.

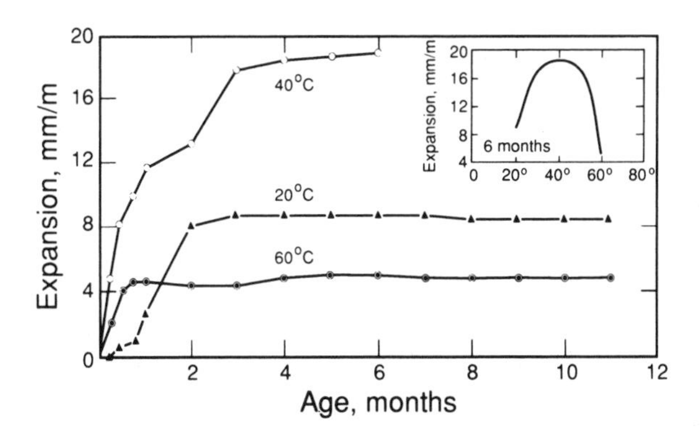

Fig. 9.12. Effect of temperature on the rate of expansion due to alkali–aggregate reaction. (Adapted from Ref. 9.26.)

9.4.2. Effect of Temperature

Temperature accelerates the rate of the alkali–aggregate reaction. This accelerating effect is demonstrated in Fig. 9.12 in which the intensity of the reaction is measured by the resulting expansion. Indeed, this effect is utilised in determining the potential alkali reactivity of cement–aggregate combinations in accordance with ASTM C227, i.e. the test in question is conducted at 37·8°C rather than at room temperature. It should be noted, however, that the effect of temperature on the expansion is characterised by a pessimum at approximately 40°C, and a further increase in temperature is associated with a lower expansion (insert in Fig. 9.12). As the damaging effect is brought about by the swelling of the alkali–silica gel on absorption of water, it is conditional on the availability of a sufficient amount of water. Hence, it may be concluded that the alkali–aggregate reactions, and their associated cracking and deterioration, will be more intensive and damaging in hot regions, or rather in hot humid regions (RH greater than, say, 85%). Much less damage, if any, is to be expected in arid zones provided, of course, the concrete is not in direct contact with water, such as may be the case in hydraulic and marine structures.

9.4.3. Controlling Alkali–Silica Reaction

It is self-evident that the alkali–silica reaction is conditional on the availability of alkalies. Consequently, unless the alkalies penetrate the concrete from an outside source (e.g. seawater), the intensity of the reaction would depend on the alkali content of the cement. That is, a lower alkalies content is expected to produce a lower expansion, and vice versa. This expected behaviour is

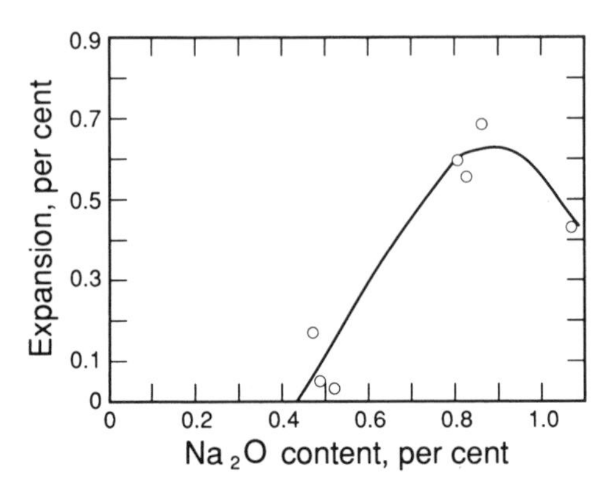

Fig. 9.13. Effect of the alkali content of the cement on the expansion of concrete due to alkali–aggregate reaction. (Adapted from Ref. 9.27.)

observed in the lower range of alkali contents, whereas a pessimum is reached at a higher content where the trend is reversed, i.e. the expansion due to the alkali-silica reaction decreases with the alkali content (Fig. 9.13). In any case, when the alkali content is low enough, i.e. approximately 0·5% of the cement by weight, in accordance with the data of Fig. 9.13, no expansion takes place. Indeed, experience has shown that no damage occurs when the total alkali content in the cement, R_2O, calculated as equivalent to Na_2O (i.e. $R_2O = Na_2O + 0{\cdot}658\,K_2O$) does not exceed 0·6%. In other words, the adverse effect of the alkali–silica reaction can be eliminated by the use of such 'low-alkalies cements'. Accordingly, this conclusion is reflected in the recommendations for the cements to be used when alkali reactive aggregates are involved (Table 9.3).

Blended cements incorporating natural pozzolan or fly-ash, or replacing Portland cement by such mineral admixtures, were shown to reduce concrete expansion due to the alkali–silica reaction. The beneficial effect of fly-ash, for example, is clearly demonstrated in Fig. 9.14 whereas similar data relevant to natural pozzolan and silica fume, can be found in Ref. 9.18 and 9.29, respectively. The exact mechanism involved is not clear as yet but, apparently, provided the Na_2O equivalent content in the concrete does not exceed $3\,kg/m^3$, the replacement of at least 25% of the cement by a pozzolan may prove to be a suitable means of controlling the alkali–silica reaction (Table 9.3). The required replacement by condensed silica fume is, apparently, much smaller [9.29].

Replacing Portland cement with granulated blast-furnace slag reduces considerably the expansion due to the alkali–aggregate reaction (Fig. 9.15). In

Table 9.3. Recommended Cements for use in Controlling Alkali–Silica Reation.[a]

No.	*Na_2O equivalent content in Portland cement per cent (by wt)*	*Recommended cement or blend*
1	0·6% or less	Portland cement
		Portland cement, blast-furnace slag cement or Portland cement blended with blast-furnace slag
2	1·1% or less	50% min slag content
	2·0% or less	65% min slag content
3	Less than 3·0 kg/m^3 Na_2O in concrete	Portland–fly-ash cement or Portland cement blended with fly-ash
		25% min fly-ash content

[a] *Adapted from Ref. 9.12.*

fact, slag cements, containing a minimum of 65% slag, were found to be suitable for controlling the alkali–aggregate reaction [9.31]. Hence, to this end, such cements can be substituted for low-alkali Portland cements (Table 9.3). The better performance of slag cements in controlling the alkali–silica reaction

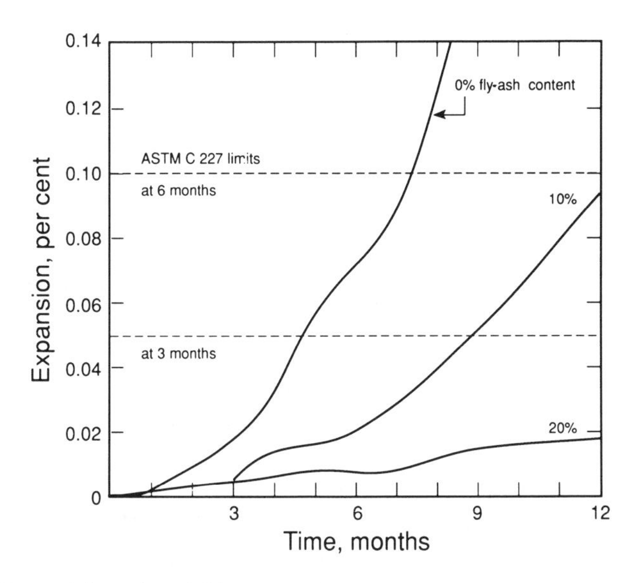

Fig. 9.14. Effect of fly-ash additions on the rate of expansion due to alkali–aggregate reaction. (Adapted from Ref. 9.28.)

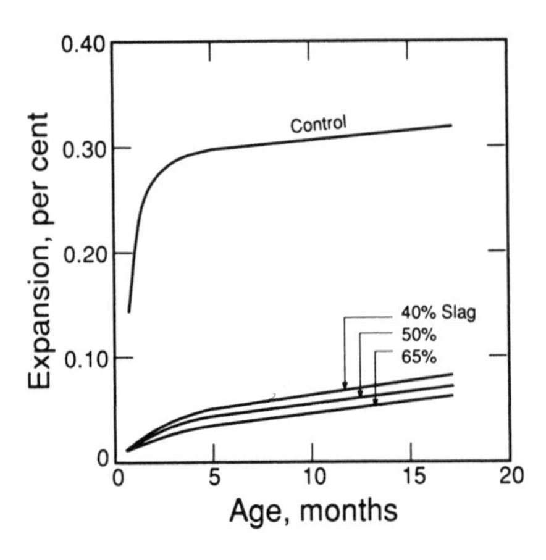

Fig. 9.15. Effect of replacing OPC with granulated blast-furnace slag on the expansion of mortars due to alkali aggregate reaction. (Adapted from Ref. 9.30.)

has been attributed to the finer pore structure and the lower permeability associated with the use of such cements (Fig. 9.3).

REFERENCES

9.1. Soroka, I., *Portland Cement Paste and Concrete*. The Macmillan Press, London, UK, 1979, pp. 145–68, 260–91.

9.2. Draft CEB guide to *Durable Concrete Structures*. Information Bull No. 166, 1985.

9.3. Soroka, I., *Portland Cement Paste and Concrete*. The Macmillan Press, London, UK, 1979, p. 88.

9.4. ACI Committee 211, Standard practice for selecting proportions for normal, heavy weight and mass concrete (ACI 211.1–89). In *ACI Manual of Concrete Practice* (Part 1). ACI, Detroit, MI, USA, 1990.

9.5. ACI Committee 201, *Guide to durable concrete*. (ACI 201.2R–77) (Reapproved 1982). In *ACI Manual of Concrete Practice* (Part 1). ACI, Detroit, MI, USA, 1990.

9.6. Goto, S. & Roy, D.M., The effect of W/C ratio and curing temperature on the permeability of hardened cement paste. *Concrete Res.*, **11**(4), (1981), 575–9.

9.7. Bakker, R.F.M., Permeability of blended cement concretes. In *Use of Fly-Ash, Silica Fume, Slag and Other Mineral By-products in Concrete* ACI Spec. Publ. SP 79, Vol. I., ed. V.M. Malhotra. ACI, Detroit, MI, USA, 1983, pp. 589–605.

9.8. Feldman, R.F., Pore structure formation during hydration of fly-ash and slag cement blends. In *Effects of Fly-Ash Incorporation in Cement and Concrete*, ed. S. Diamond. Materials Research Society, PA, USA, 1981, pp. 124–33.

9.9. Manmohan, D. & Mehta, P.K., Influence of pozzolanic, slag and chemical admixtures on pore size distribution and permeability of hardened cement pastes. *Cement, Concrete and Aggregates*, **3**(1), 1981, 63–67.

9.10. Sellevold, E.J., Baker, D.H., Jensen, E.K. & Knudsen, T., Silica fume cement paste – hydration and pore structure. In *Condensed Silica Fume in Concrete*. Norwegian Inst. of Technology, Univ. of Trondheim, Norway, Report BML 82–610, Feb. 1982, pp. 19–50.
9.11. Elola, A.I., Szteinberg, A.S. & Torrent, R.J., Effect of the addition of blast-furnace slag on the physical and mechanical properties of mortar cured at high temperatures. In *Proc. Symp. Chem. Cement*, Vol. 4, 1986, Sindicato Nacional da Industria do Cimento, Rio de Janeiro, pp. 145–9.
9.12. STUVO, *Concrete in Hot Countries*. The Dutch member group of FIP, The Netherlands.
9.13. Mather, B., Field and laboratory studies of the sulphate resistance of concrete. In *Performance of Concrete*, ed. G.E. Swenson. University of Toronto Press, Toronto, Canada, 1968, pp. 66–76.
9.14. Verbeck, G.J., Field and laboratory studies of the sulphate resistance of concrete. In *Performance of Concrete*, ed. G.E. Swenson. University of Toronto Press, Toronto, Canada, 1968, pp. 113–24.
9.15. Brown G.E. & Oates, D.B., Air entrainment in sulfate-resistant concrete. *Concrete Int.*, **5**(1) (1983), 36–9.
9.16. Cabrera, J. & Plowman, C., The mechanism and rate of attack of sodium sulphate solution on cement and cement pfa pastes. *Adv. Cement Res.*, **1**(3) (1988), 171–9.
9.17. Dunstan, E.R., A possible method for identifying fly ashes that will improve sulphate resistance of concretes. *Cement, Concrete and Aggregates*, **2**(1) (1980), 20–30.
9.18. Mehta, P.K., Studies on blended Portland cements containing Santorin earth. *Cement Concrete Res.*, **11**(4) (1981), 507–18.
9.19. Dunstan, E.R., Performance of lignite and sub-bituminous fly ash in concrete – A progress report. Report REC-ERC-76-1, US Bureau of Reclamation, Denver, CO, USA, 1976.
9.20. Locher, F.W., The problem of the sulphate resistance of slag cements. *Zement–Kalk–Gips*, **19**(9) (1966), 395–401 (in German).
9.21. Ludwig, M. & Darr, G.J., On the sulphate resistance of cement mortars. Research Report of the States of Nordheim–Westfalen No. 2636, 1976 (in German).
9.22. Lea, F.M., *The Chemistry of Cement and Concrete*. Edward Arnold, London, UK, 1970, p. 348.
9.23. Biczok, I., *Concrete Corrosion – Concrete Protection*. Akademiai Kiado, Budapest, Hungary, 1972, p. 217.
9.24. Locher, F.W., Influence of chloride and hydrocarbonate on sulphate attack. In *Proc. Symp. Chem. of Cement*, Tokyo, 1968, Vol. 3, The Cement Association of Japan, Tokyo, pp. 328–35.

9.25. Lea, F.M., *The Chemistry of Cement and Concrete*. Edward Arnold, London, UK, 1970, pp. 439–43.

9.26. Locher, F.W. & Sprung, S., Origin and nature of alkali–aggregate reaction. *Beton*, **23**(7) (1973), 303–6 (in German).

9.27. Woods, H., Durability of concrete construction. Monograph No. 4, ACI, Detroit, MI, USA, 1968.

9.28 Stark, D., Alkali silica reactivity in the Rocky Mountain region. In *Proc. 4th Intern. Conf. Effects of Alkalies in Cement and Concrete*. Purdue University, W. Lafayette, IN, USA, 1978, pp. 235–43.

9.29. Sellevold, E.J. & Nilsen, T., Condensed silica fume in concrete: A world review. In *Supplementary Cementing Materials for Concrete*, ed. V.M. Malhotra. CANMET, Ottawa, Canada, 1987, pp. 167–229.

9.30. Hogan, F.J. & Meusel, J.W., Evaluation for durability and strength development of ground granulated blastfurnace slag. *Cement, Concrete and Aggregates*, **3**(1) (1981), 40–52.

9.31. Bakker, R.F.M., On the causes of increased resistance of concrete made of blast-furnace slag cement to alkali–silica reaction and to sulphate corrosion. Doctorate Thesis, T.H. Aachen, Germany, 1980 (in German).

Chapter 10
Corrosion of Reinforcement

10.1. INTRODUCTION

The formation of the corrosion products of iron (i.e. rust) involves a substantial volume increase, i.e. the volume of the corrosion products, assuming they are mainly $Fe(OH)_3$, is some four times greater than that of the corroding iron. In reinforced concrete, such an expansion is subjected to volume restraint and, therefore, when rust is formed, pressure is exerted on the surrounding concrete. At some stage, this pressure may cause the cracking of the concrete cover over the reinforcement, and the corrosion is then aggravated due to the readily available oxygen and moisture, which are conditional for the corrosion process to proceed. At a more advanced stage, spalling of the concrete cover occurs, and the unprotected reinforcement is exposed to environmental factors. The continued corrosion of the reinforcement gradually reduces the cross-sectional area of the reinforcing bars (i.e. rebars) and thereby, also, the bearing capacity of the structural element involved. Hence, if no remedial means are employed, and depending on the severity of the exposure conditions, complete deterioration and failure may follow, and the end of the structure, the so-called 'service-life', is reached.

The service-life of a reinforced concrete structure, with respect to the corrosion process, is schematically described in Fig. 10.1. In the first stage, usually referred to as the 'initiation' or the 'incubation' period, no corrosion occurs because the rebars are protected by the high alkalinity of the concrete (see section 10.3). This period lasts until the concrete carbonates to the depth

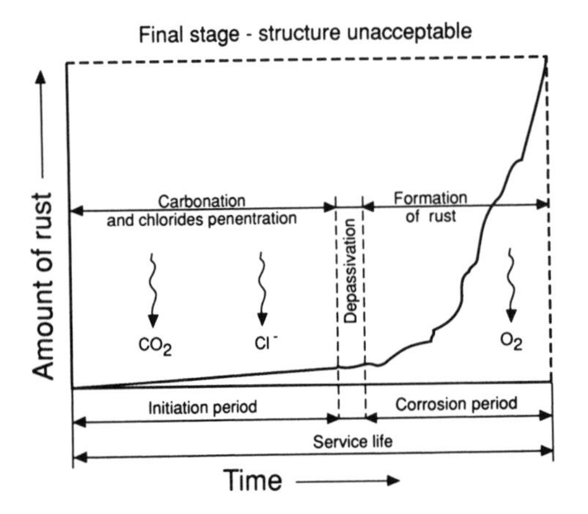

Fig. 10.1. Schematic description of the service-life of a reinforced concrete structure.

of the rebars (see section 10.4) or the chloride content at the level of the rebars reaches a critical value (see section 10.5). When such a stage is reached, a so-called 'depassivation' process occurs (see section 10.3), and corrosion takes place provided both moisture and oxygen are available. As explained earlier, the formation of rust involves the development of disruptive pressure, which subsequently may lead to cracking and spalling. The corrosion propagates at a rate which depends on the rate of oxygen supply at the rebars level and at a certain stage, unless the concrete is repaired, the element involved becomes unsafe and unusable any longer, i.e. the final stage of the structure's service-life is reached. It may be realised that not only the integrity of the structure determines its service life, and it may become unacceptable due to extensive cracking and spalling, and even due to excessive rust-staining.

It is obvious from the preceding discussion that corrosion of the reinforcement may adversely and significantly affect the durability of concrete structures. In fact, corrosion of reinforcement is, by far, the most damaging process with respect to the durability of concrete structures. Hence, the need to protect the reinforcement against corrosion cannot be overestimated, and such protection must always be provided. Moreover, as will be seen later, elevated temperatures and the presence of chlorides both aggravate the situation. Hence, when such conditions prevail, extra care must be exercised in protecting the reinforcing steel. The factors which affect the corrosion process, as well as suitable means for protecting the reinforcement against such a process, are discussed below in some detail.

10.2. MECHANISM

Generally, corrosion is the deterioration and the slow wearing away of solids, especially metals, by chemical attack. The most common type of such corrosion involves metal oxidation which is brought about by an electrochemical process. The mechanism of the latter process may be somewhat complicated. In the following discussion, however, it is treated in a simplified way and with particular reference to iron.

When iron is placed in water, the latter goes into solution as positively charged ions, and negatively charged electrons are released:

$$Fe \rightleftarrows Fe^{2+} + 2e^- \qquad (10.1)$$

$$F^{2+} \rightleftarrows Fe^{3+} + e^- \qquad (10.2)$$

As a result of the reactions, an electrical potential, known as 'electrode potential', is built up. It follows that the higher the solubility of the metal, the higher the electrode potential and the greater the corroding tendency of the metal involved. In turn, the solubility, as such, depends on the nature of the metal in question, and on that of the solution. In this respect, it is of interest to note that the presence of chloride ions in the solution increases the solubility of iron and thereby its vulnerability to corrosion.

It may be noted that the reactions in eqns (10.1) and (10.2) are reversible and equilibrium is reached rather quickly. The latter reactions, however, will continue if the electrons produced are removed from the iron, and thereby prevent equilibrium from being reached. Indeed, this happens to be the case when the iron electrode is connected to a metal electrode of a lower potential and oxygen is available at the latter electrode. Under such conditions, the electrons flowing from the iron electrode are consumed in accordance with the following expression:

$$2H_2O + O_2 + 4e^- \rightarrow 4(OH)^- \qquad (10.3)$$

The electrode of the higher potential, i.e. the electrode producing the electrons, is called the 'anode' and the one at which the electrons are consumed is called the 'cathode'. The Fe^{3+} ions, which are produced at the anode (eqn (10.2)), diffuse towards the cathode and combine with the hydroxyl ions to give rust:

$$Fe^{3+} + 3(OH)^- \rightarrow Fe(OH)_3 \text{ (rust)} \qquad (10.4)$$

The Fe^{3+} ions are much smaller than the hydroxyl ions and, therefore, the Fe^{3+} ions diffuse more rapidly than the OH^- ions. Furthermore, only one F^{3+} ion

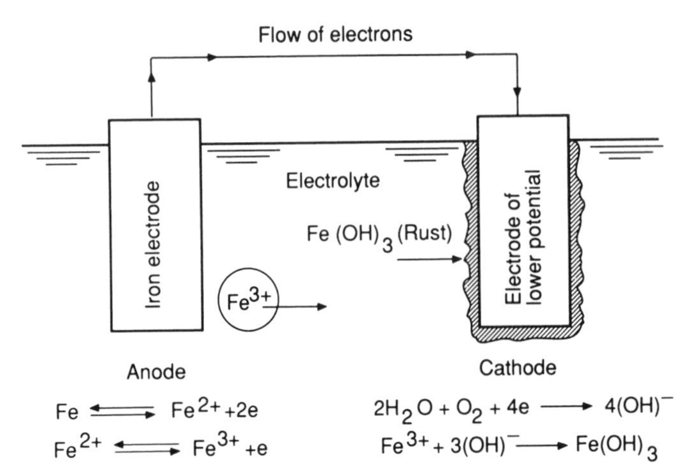

Fig. 10.2. Schematic description of iron corrosion due to electrochemical process.

is required to combine with three OH^- ions. Consequently, eqn (10.4) mostly occurs at the cathode. In other words, corrosion occurs at the anode, whereas the rust is deposited mainly at the cathode (Fig. 10.2).

In view of the preceding discussion, it is clear that the corrosion process is conditional on the presence of both water and oxygen. The water constitutes an electrolyte which facilitates the diffusion of the Fe^{3+} ions from the anode to the cathode, whereas the oxygen is required for the consumption of the electrons. It also increases the hydroxyl ion supply, which is needed for the rust forming reaction (eqn (10.4)). Hence, because of the lack of oxygen, no corrosion is expected in iron which is completely submerged in water. Similarly, no corrosion is expected in a very dry environment. In practice, however, only little moisture is required to promote corrosion.

It must be realised that the corrosion process is not, necessarily, conditional on the contact between two dissimilar metals. The formation of anodic and cathodic sites may occur in the same metal, due to local variations in its composition, stress level, oxygen supply, etc. Hence, anodic and cathodic sites develop, sometimes, at very short intervals, to give what is usually referred to as galvanic microcells. This may be the case, for example, in steel, which contains iron and carbide (FeC_3). The carbide, due to its lower electric potential, constitutes a cathodic site, and thereby brings about the corrosion of the iron of the steel (Fig. 10.3).

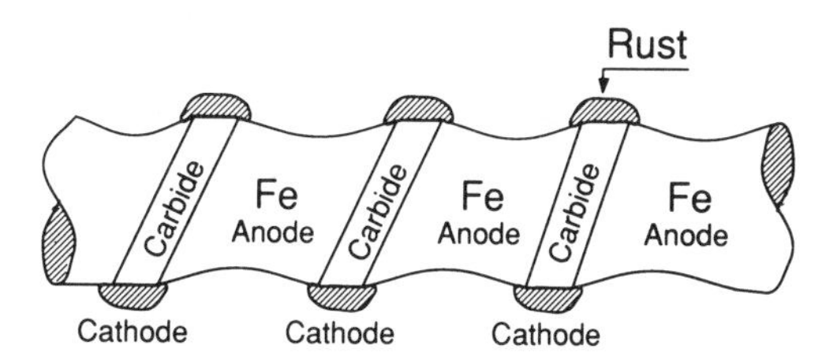

Fig. 10.3. The formation of galvanic microcells in steel due to the presence of carbide.

10.3. CORROSION OF STEEL IN CONCRETE

Concrete protects the embedded steel reinforcement against corrosion due to the high alkalinity of the pore water of the cement paste. The pH of the pore water varies from 12·5 to 13·5, and under such conditions a thin oxide layer is formed on the surface of the rebars and prevents the iron from dissolving, i.e. corrosion is prevented even in the presence of moisture and oxygen. This protective film is referred to as the 'passive film', and its protective effect on the steel against corrosion, as 'passivation'. It follows that, as long as the passive film remains intact, the rebars remain protected from corrosion. This is the case when the pH of the pore water in contact with the rebars exceeds, say, 9. At lower pH levels, a 'depassivation' process occurs (i.e. the passive film disintegrates) which leaves the steel unprotected and prone to corrosion. Similarly, depassivation may occur due to the presence of chloride ions at the rebars level. Hence, maintaining the pH level of the pore water greater than 9, and preventing the chlorides from reaching the rebars level, would prevent corrosion of the latter. Indeed, in practice, such prevention is achieved by providing the rebars with a dense concrete cover of adequate thickness.

10.4. CARBONATION

The high alkalinity of concrete is partly due to the presence of the alkalis Na_2O and K_2O of the cement, and mainly to the presence of calcium hydroxide which is produced on the hydration of the Alite and the Belite (see section 2.3). Normal air contains some 0·03% carbon dioxide (CO_2) by volume. The capillary pore system of the cement paste allows air to penetrate into the concrete and the CO_2 of the air combines with the calcium hydroxide ($Ca(OH)_2$) to give calcium carbonate ($CaCO_3$) in accordance with the following expression:

$$Ca(OH)_2 + CO_2 \rightarrow CaCO_3 + H_2O \quad (10.5)$$

The transformation of the $Ca(OH)_2$ to $CaCO_3$, referred to as 'carbonation', lowers the pH of the pore water to less than 9 in a fully carbonated concrete. Once this pH level is reached at the surface of the rebars, depassivation occurs and the onset of corrosion takes place provided, as it is usually the case, moisture and oxygen are available. A schematic description of the carbonation of concrete and the resulting corrosion of the rebars is presented in Fig. 10.4.

Carbonation starts at the concrete surface and proceeds inward at a rate which depends on concrete quality (i.e. mainly its porosity) on the one hand, and environmental factors such as humidity temperature, on the other. Generally, the relation between the depth of carbonation, d, and the time, t, it takes the carbonation to reach such a depth, is given by the following expression:

$$d = k \times t^{1/n} \quad (10.6)$$

where k is a constant which depends on all factors which determine the rate of carbonation (i.e. concrete quality and environmental conditions).

The value of $1/n$ varies from, say, 0·35 to 0·65, but 0·5 is sometimes considered as an approximate average. The latter expression may be used to estimate the minimum thickness of the concrete cover which is required, at the conditions considered, for the carbonation front to reach the reinforcement level at a given time. This time is very sensitive to the thickness of the reinforcement cover, i.e. doubling the thickness increases the carbonation time by a factor of four, when $1/n$ is assumed to equal 0·5. In other words, cover thickness constitutes an efficient means to control carbonation, and thereby also the onset of corrosion of the reinforcing steel. Controlling the onset of the

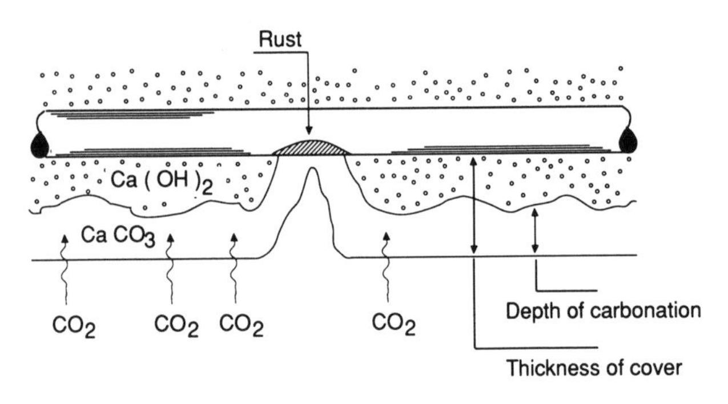

Fig. 10.4. Schematic description of the carbonation process.

corrosion process is very important in giving the required 'service life' to reinforced concrete, and this specific aspect is discussed below (see sections 10.4.1.2 and 10.5.1.1).

10.4.1. Factors Affecting Rate of Carbonation

It is self-evident that the rate of carbonation is determined by the rate of CO_2 diffusion into the concrete. In turn, this diffusion depends on concrete porosity and its moisture content. Hence, the rate of carbonation depends on the very same factors.

10.4.1.1. Environmental Conditions

The diffusion of CO_2 in water is very low indeed and, accordingly, hardly any carbonation takes place in a fully saturated concrete. On the other hand, the presence of some moisture is necessary to allow carbonation to proceed. Hence, no carbonation occurs when the concrete is completely dry. It follows that an optimum moisture content, at which carbonation proceeds at a maximum rate, must exist between the two extremes. Indeed, it is generally accepted that such a maximum is reached when the concrete is exposed to the relative humidity of 50–60% (Fig. 10.5).

It may be noted that in accordance with the data of Fig. 10.5, the effect of temperature on carbonation in the range 5–20°C, if any, is very small.

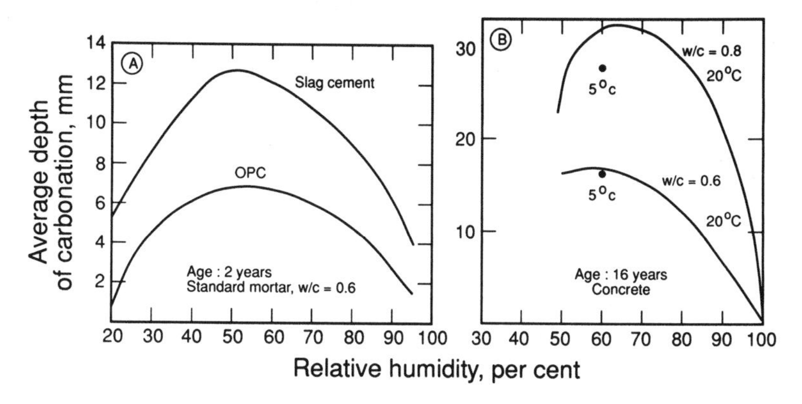

Fig. 10.5. Effect of relative humidity on carbonation of mortar and concrete. (A) Standard cement mortar (water to cement ratio (W/C) = 0·6) at the age of 2 years. (B) Concrete (W/C ratios 0·6 and 0·8, temperature 20°C) at the age of 16 years. (Adapted from Ref. 10.1.)

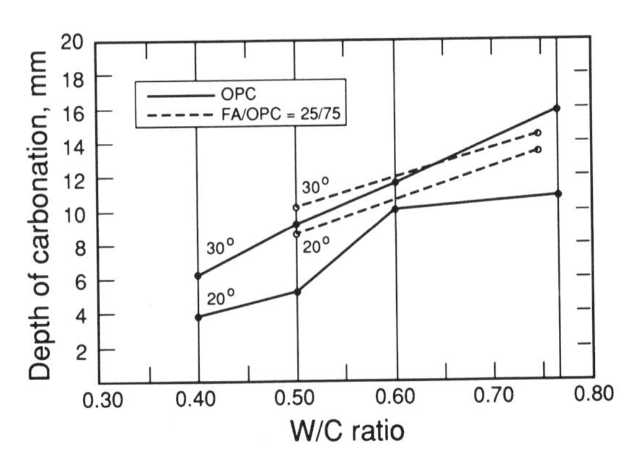

Fig. 10.6. Effect of W/C ratio, type of cement and temperature on the depth of carbonation of concrete at the age of 15 months. (Adapted from Ref. 10.2.)

(Compare points marked 5°C with curves in part B of Fig. 10.5.) Figure 10.6, however indicates that carbonation increases with temperature in the range of 20–30°C. A similar increase was observed by others [10.3] and it is usually accepted that carbonation increases with temperature. It must be realised that this conclusion may be considered valid for the effect of temperature *per se*, whereas in practice a higher temperature may involve a more intensive drying and, thereby, bring about a lower rate of carbonation due to the reduced moisture content. That is, an increase in temperature, and the simultaneous decrease in moisture content, involve two opposing effects which may result in decreased carbonation. In fact, the possibility of such a decrease has been recognised [10.4] but, apparently, in practice this rarely occurs.

It is well recognised that concrete carbonates at a slower rate when exposed outdoors than when stored under constant laboratory conditions. Furthermore, less carbonation occurs in concrete which is exposed outdoors unprotected from precipitation than in the same concrete which is protected from getting wet (Fig. 10.7). This effect of environmental conditions is attributable to the effects of moisture content and temperature on the rate of carbonation. The wetting effect of the rain and the lower outdoor temperatures both reduce the rate of carbonation, explaining, in turn, the lower carbonation of concrete which is exposed to outdoor conditions. This behaviour, which is observed in mild and cold regions, is not necessarily expected in hot regions where laboratory storage may involve lower temperatures, and in hot, dry regions, where laboratory storage may also involve higher relative humidities than those prevailing outdoors. In any case, predicting the exact

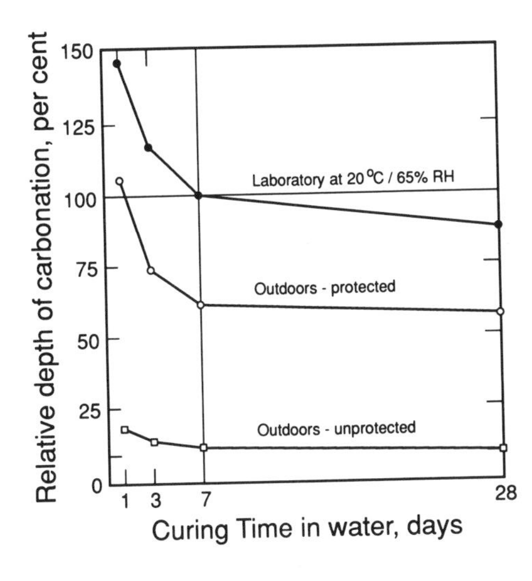

Fig. 10.7. Effect of curing time and exposure conditions on depth of carbonation of concrete at the age of 16 years. (Adapted from Ref. 10.1.)

effect of specific environmental conditions on the carbonation of concrete is rather difficult, if not impossible.

10.4.1.2. Porosity of Concrete Cover

The porosity of concrete is determined by the W/C ratio and the degree of hydration, whereas the degree of hydration is determined by the length of curing time and its effectiveness. That is, the lower the W/C ratio and the longer the curing period, the lower the porosity of the concrete. Hence, it is to be expected that the rate of carbonation will be affected similarly, i.e. the rate will increase with an increase in the W/C ratio and a decrease in the length of curing time. This expected effect of the W/C ratio is demonstrated in Figs 10.6 and 10.8, and that of the length of curing period in Figs 10.7 and 10.8. In this respect, it may be noted from Fig. 10.7 that, at least for the conditions considered, curing periods longer than 7 days hardly affect significantly the depth of carbonation.

10.4.1.3. Type of Cement and Cement Content

The high alkalinity of the pore water in concrete is mainly due to the presence of calcium hydroxide. Hence, under otherwise the same conditions, the rate of carbonation is expected to decrease with an increase in the calcium hydroxide content. The calcium hydroxide originates in the hydration of both the Alite (C_3S) and the Belite (C_2S) of the cement, but a much greater quantity is produced from the hydration of the C_3S than from the hydration of the C_2S (see section 2.3). In other words, the calcium hydroxide content in concrete

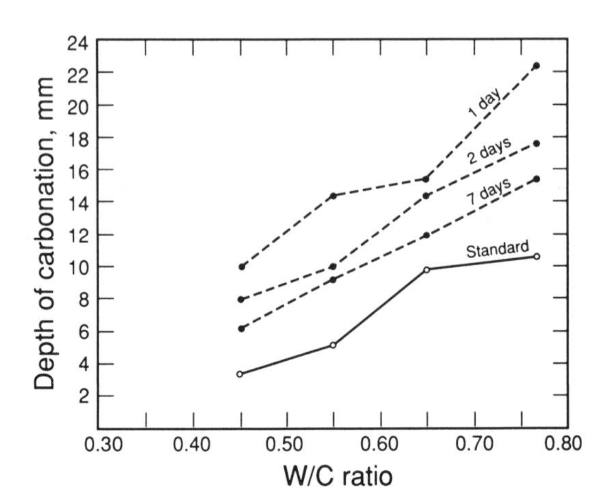

Fig. 10.8. Effect of W/C ratio and length of curing at 30°C on depth of carbonation at the age of 15 months. (Adapted from Ref. 10.2.)

made of C_3S-rich cements, such as rapid-hardening cements, is greater than in that made of ordinary or low-heat Portland cements, and, consequently, the former cements are expected to impart the concrete a lower rate of carbonation. Indeed, this lower rate of carbonation was observed in mortars made of rapid-hardening cement (Fig. 10.9). On the other hand, for the very same reasons, a cement rich in C_2S, such as low-heat Portland cement, and sometimes also sulphate-resisting Portland cement, is expected to exhibit a higher rate of carbonation than ordinary Portland cement (OPC). This, however, is not necessarily always the case (Fig. 10.9).

An increase in the calcium hydroxide content is also brought about by an increase in the cement content. Accordingly, a lower rate of carbonation is to be expected in cement-rich concretes than in their lean counterparts. This effect of the cement content is attributable, not only to its effect on the calcium

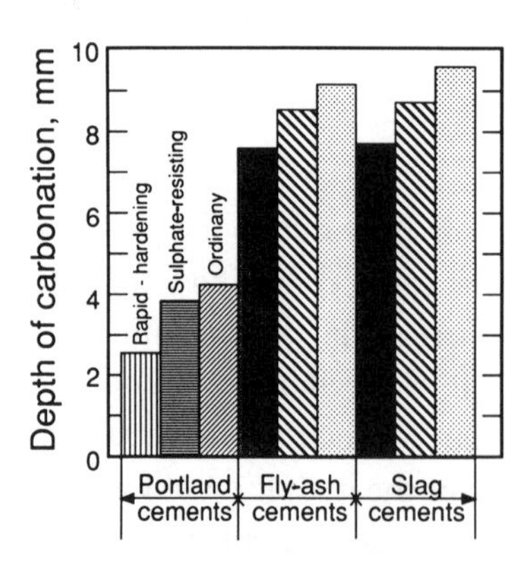

Fig. 10.9. Effect of type of cement on depth of carbonation of mortars at the age of 6 months (W/C = 0·65). (Adapted from Ref. 10.5.)

hydroxide content, but also to its effect on the W/C ratio. In practice, an increased cement content is usually associated with a decreased W/C ratio, whereas, as pointed out earlier (Figs 10.6 and 10.8), a decreased W/C ratio involves a lower rate of carbonation.

In the carbonation of concretes made of blended cements, two opposing effects are involved explaining, in turn, the sometimes contradictory nature of the reported data. The use of blended cements involves a reduced calcium hydroxide content due, in the first instance, to the diluting effect of the admixture and, in some cases, also due to pozzolanic reactions which further reduce the calcium hydroxide content of the concrete. That is, a higher rate of carbonation may be expected in concretes made of blended cements and particularly in those made of Portland–pozzolan cements. On the other hand, the latter concretes are characterised by a finer pore system (Chapter 3, Figs 3.3 and 3.15) and a lower permeability (Chapter 9, Fig. 9.3), and should exhibit, therefore, a lower rate of carbonation. Apparently, the net effect of these two opposing factors is negative and blended cement concretes usually exhibit a higher rate of carbonation than otherwise the same concretes made of ordinary Portland cement. This higher rate of carbonation is demonstrated in Figs 10.5(A) and 10.9 for slag cement concrete, and in Fig. 10.9 for fly-ash concrete. It may be noted that this effect of fly-ash is hardly reflected in Fig. 10.6 in which the depth of carbonation at 30°C, for example, is virtually the same for ordinary and fly-ash concretes. Again, this specific observation may be attributed to the opposing effects involved but, as pointed out earlier, the use of blended cements is usually associated with a higher rate of carbonation.

10.4.1.4. Practical Conclusions

In order to reduce the rate of carbonation and the time it takes the carbonation front to reach the rebars level, and thereby cause depassivation and subsequent corrosion, the reinforcing steel must be provided with a dense concrete cover of adequate thickness. A low W/C ratio and proper curing are required to yield a concrete with the desired density.

When carbonation only is considered, the use of OPC is preferable to blended cements. It must be realised, however, that due to the many factors involved, this conclusion is not necessarily valid when the corrosion of the reinforcement is considered. This aspect of the more suitable cement, from the corrosion point of view, is treated below (see section 10.8).

10.5. CHLORIDE PENETRATION

It was mentioned earlier that depassivation may take place due to the presence of chloride ions at the reinforcement level. Chloride ions may be present in the concrete due to the use of contaminated aggregates or chloride-containing admixtures, or due to penetration from external sources such as seawater or a marine environment. Another source may be de-icing salts which are used to prevent frost-damage to concrete. This latter source of chlorides is, of course, not relevant to the subject at hand.

Unlike the diffusion of the CO_2, that of the chloride ions takes place only in water. Hence, chloride penetration is conditional on the presence of water in the pore system. The mechanism involved is either capillary suction of chloride-containing water, or simply diffusion of ions in the still pore water. In other words, in the first case, which is characteristic mainly of comparatively dry concrete, the water constitutes a vehicle which carries the ions into the concrete. In the second case, which is characteristic mainly of saturated or nearly saturated concrete, the water constitutes a medium through which the ions diffuse inside. In concrete which is exposed to alternate cycles of wetting and drying, both mechanisms are operative and therefore, under such conditions, an increased rate of chlorides penetration is to be expected.

A typical chloride penetration profile is presented in Fig. 10.10. Generally speaking, chloride concentration increases with time and, as can be expected, decreases with the distance from the concrete surface, i.e. with the depth of

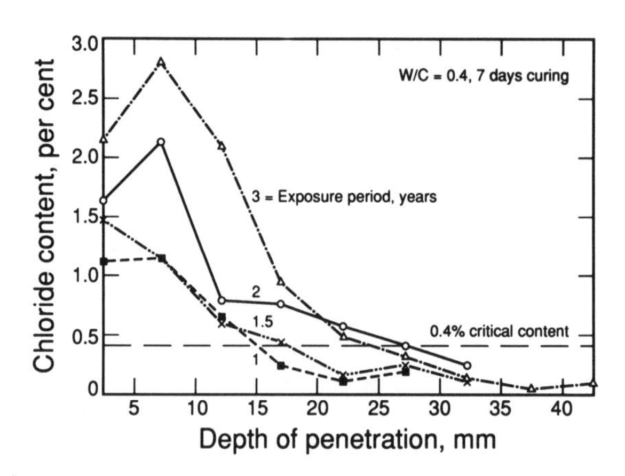

Fig. 10.10. Effect of exposure time on profiles of chloride penetration into concrete (W/C ratio = 0·40, 7 days curing). (Adapted from Ref. 10.7.)

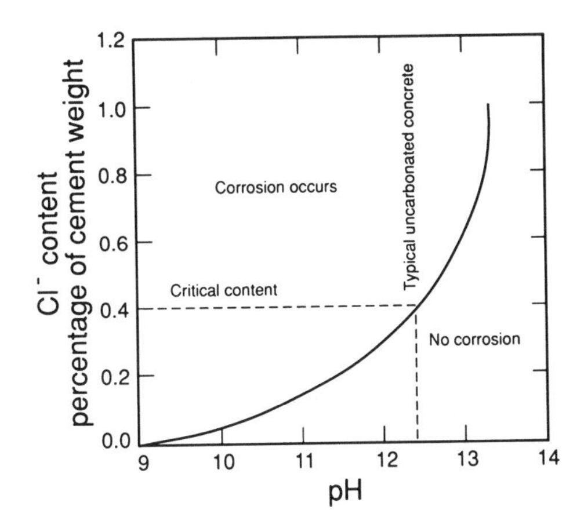

Fig. 10.11. The relationship between the pH of the pore water and critical chloride content.

penetration. In order to prevent the depassivation of the rebars, the chloride ions must be prevented from reaching the reinforcement and exceed a certain critical concentration. Again, this can be achieved by providing the rebars with a dense concrete cover of adequate thickness, i.e. by the very same means which are suitable to control carbonation.

As mentioned earlier, in order to cause depassivation, the chloride concentration must exceed a certain level. As will be explained later (see section 10.5.1.2), some of the chlorides which penetrate into the concrete combine with the alumina-bearing phases of the hydration products, whereas only the free chlorides may cause depassivation. Usually, in the analysis of chloride content, the total is determined and, therefore, this total must exceed the amount of the combined chlorides in order to cause depassivation. The latter content, usually referred to as the 'critical' or the 'threshold' content, depends on the pH value of the pore water and is usually expressed as a percentage of the cement weight. The critical content depends also on some additional factors, such as the cement composition. It is generally assumed that, however, for uncarbonated concrete, the critical content is 0·4% of the cement weight, decreasing to zero for pH = 9, i.e. for the pH for which depassivation occurs anyway due to the low alkalinity of the pore water. This relation between the pH of the pore water and critical chloride content is schematically described in Fig. 10.11.

10.5.1. Factors Affecting Rate of Chloride Penetration

10.5.1.1. Porosity of Concrete Cover

The time it takes chloride concentration to reach the critical content of 0·4% at a certain distance from the surface, increases with the decrease in concrete

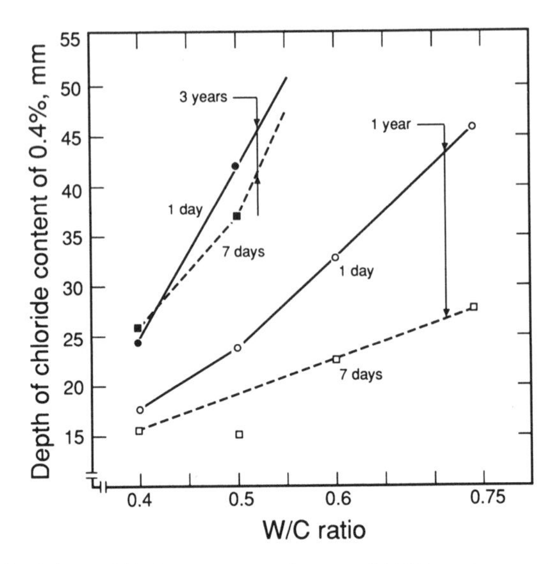

Fig. 10.12. Effect of 1 and 7 days wet curing and W/C ratio on the depth at which the chloride content reached the critical value of 0·4% after 1 and 3 years' exposure. (Adapted from Ref. 10.7.)

porosity or, alternatively, the distance at which the critical content is reached at a given time, increases with the increase in concrete porosity. It follows that the W/C ratio and length of curing would affect, similarly, chloride penetration [10.6]. This expected effect, with some exception with respect to the effect of curing at the age of 3 years, is demonstrated in the data of Fig. 10.12. Accordingly, it may be again concluded that in order to control chloride penetration, a well-cured concrete cover, made with a low W/C ratio, should be provided over the rebars.

10.5.1.2. Type of Cement and Cement Content

It was pointed out earlier that, with respect to depassivation, only the free chlorides are important, i.e. only the chlorides which are not bound by, or adsorbed on, the hydration products. The adsorption of the chlorides is less clear, but the chlorides combine mostly with the C_3A hydration products to give Friedles salt ($3CaO.Al_2O_3.CaCl_2.10H_2O$) or, when concentrated chloride solutions are involved, also calcium oxychloride ($CaO.CaCl_2.2H_2O$). It follows that the binding capacity of Portland cement is determined by its C_3A content and, in this context, a high C_3A content cement is, therefore, to be preferred. It may be expected that such a cement, due to its greater binding capacity, will slow down the ingress of chlorides into the concrete, and thereby

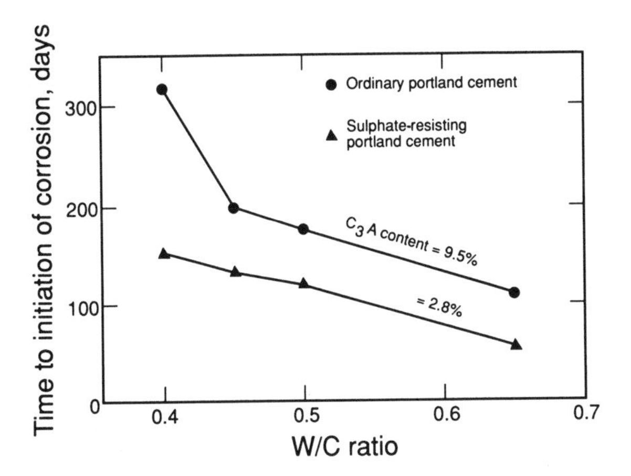

Fig. 10.13. Effect of C_3A content of Portland cement on the length of corrosion initiation period in concrete specimens partially immersed in 5% sodium chloride solution. (Adapted from Ref. 10.8.)

bring about a longer initiation period (Fig. 10.1). This effect of the C_3A content on the initiation period is demonstrated in Fig. 10.13 and, accordingly, it may be concluded that in this respect, OPC, because of its higher C_3A content, is preferable to its sulphate-resisting counterpart.

It is self-evident that the amount of the combined chlorides is determined, not only by the cement composition, but also by the cement content in the concrete. It is to be expected that, accordingly, the higher the cement content the greater the amount of the chlorides which are combined, and the slower the rate of chloride penetration. This effect of the cement content is confirmed by the data of Fig. 10.14 and, indeed, relating the critical chloride content to the weight of the cement, recognises this effect of the cement content.

In considering blended cements with respect to their binding capacity of chloride ions, distinction should be made between pozzolanic and slag cements.

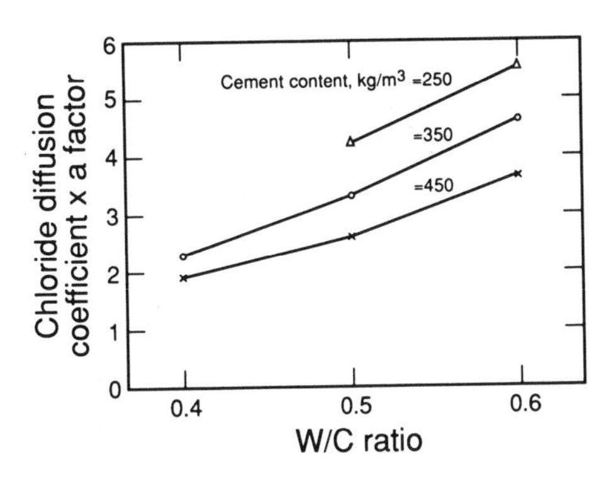

Fig. 10.14. Effect of W/C ratio and the cement content on chloride diffusion coefficient in concrete specimens placed in intertidal range of seawater. (Adapted from Ref. 10.9.)

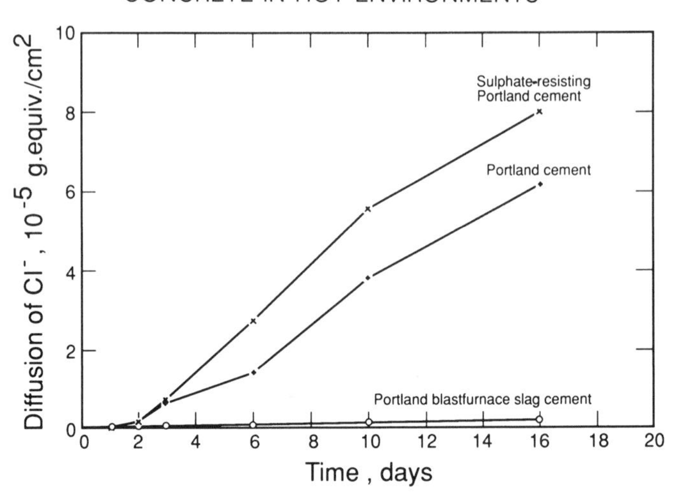

Fig. 10.15. Effect of type of cement on the rate of diffusion of chloride ions. (Adapted from Ref. 10.10.)

The hydration of blast-furnace slag gives a hydrate of calcium aluminate ($4CaO.Al_2O.nH_2O$) and the binding capacity of the slag cement is determined, therefore, by both the C_3A content of the Portland cement and the alumina content of the slag. No aluminates are usually produced as a result of the pozzolanic reaction and the binding capacity of pozzolanic cement is determined, therefore, only by the C_3A content of the Portland cement. Hence, considering chloride binding capacity, OPC and slag cements are preferable to pozzolanic cements.

The preceding observation that when the chloride binding capacity is considered, OPC is preferable to pozzolanic cements must not be interpreted to mean that the former cement is preferable when chloride-induced corrosion is expected. In fact, the opposite may be concluded when the rate of chloride penetration, rather than the chloride-binding capacity, is considered. It has been demonstrated that the finer porosity of blended cements is associated, not only with reduced permeability, but also with a considerably reduced rate of chloride diffusion. This reduced rate of diffusion is demonstrated, for example, for blast-furnace slag cement in Fig. 10.15, which indicates that the use of the latter cement virtually prevents the penetration of the chloride ions into the concrete. The same effect, essentially, was observed in fly-ash cements [10.11], or when condensed silica fume constituted the blending component [10.12]. That is, considering the rate of chloride penetration, the use of blended cements is preferable to that of OPC.

In practice, the rate of chloride penetration, rather than the chloride-binding capacity, should be considered in selecting the more suitable type of cement.

Indeed, it was found that the C_3A content of the cement (i.e. its binding capacity) is not as important as may be implied, and in this respect the fineness of the pore system of the cement paste is more important [10.13]. Accordingly, in terms of corrosion, the use of blended cements is preferable regardless of their possible lower binding capacity. Still, it may be further questioned which type of blended cement is preferable, i.e. slag or, say, fly-ash cement. Apparently, when elevated temperatures are involved, slag cement is preferable because the permeability of the fly-ash cement increases with the rise in temperature whereas that of slag cement remains unchanged (Chapter 9, Fig. 9.3). Hence, the use of slag cement (minimum slag content of 65%) is sometimes recommended when chloride induced corrosion is to be expected, i.e. in marine environments, etc.

10.5.1.3. Temperature

A diffusion process, including that of chlorides, usually follows the Arrhenius equation (see section 2.5.1). Accordingly, the diffusion rate of the chlorides is expected to increase with temperature, and the relation between the logarithm of the diffusion coefficient and the reciprocal of temperature, expressed in K, is expected to be linear. This is supported by the data of Fig. 10.16, implying that in a hot environment the time it takes the chlorides to reach the rebars level is shorter than in a moderate environment. Again, it may be noted, from Fig. 10.16, that the rate of chloride penetration decreases with a decrease in the

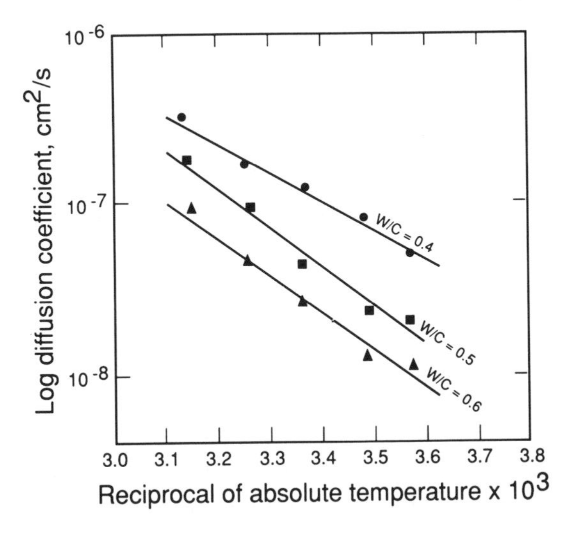

Fig. 10.16. Effect of temperature and W/C ratio on chlorides diffusion coefficient. (Adapted from Ref. 10.14.)

W/C ratio. This aspect of the W/C ratio is discussed earlier in the text (see section 10.5.1.1).

10.5.1.4. Corrosion Inhibitors

In the last two or three decades, some admixtures have been suggested to counteract the depassivation effect of the chlorides and thereby prevent or, at least delay, the corrosion of the rebars. A few types of admixtures have been considered but, apparently, the more promising ones are the nitrites, namely calcium nitrites ($Ca(NO_2)_2$) and sodium nitrites ($NaNO_2$).

On destabilisation of the passivation film, or due to penetration of chlorides through the already existing defects in the film, a soluble complex of iron chloride is formed. The resulting complex dissolves, moves away from the rebar into the concrete, and eventually precipitates as rust in accordance with the following equation:

$$[FeCl]^+ + (OH)^- \rightarrow Fe(OH)_2 + Cl^- \tag{10.7}$$

The resulting free chlorides diffuse back to the rebar and the process is repeated, i.e. a continued corrosion takes place. The presence of nitrites (NO_2) inhibits this process because the latter immediately react with the Fe^{2+} ions and ferric oxide (Fe_2O_3) is produced at the rebars' surface:

$$2Fe^{2+} 2NO_2^- + H_2O \rightarrow Fe_2O_3 + 2NO + 2H^+ \tag{10.8}$$

The resulting ferric oxide increases the thickness of the passivation layer, and thereby counteracts the depassivation effect of the chlorides. Accordingly, once the nitrite concentration is high enough with respect to that of the chlorides, no depassivation and, consequently, no corrosion, are to be expected. It is claimed that, to this end, the Cl to NO_2 ratio should be lower than 1·5–2·0. Hence, the addition of 2% of nitrite by the weight of the cement is usually recommended, and it is suggested that such an addition is enough to inhibit the corrosion of the rebars, provided the chloride content in the concrete does not exceed 10 kg/m^3, i.e. some 3% of the cement weight in normal concrete. This is, of course, a much greater percentage than the 0·4% which is considered the critical value in normal concrete containing no inhibitors.

It must be realised that there exists some uncertainty about the effectiveness of inhibitors, particularly about their long-time effect. Apparently, at best, the use of inhibitors delays corrosion rather than prevents it indefinitely. Moreover, it is sometimes argued that the use of low additions may be harmful because at low concentration the nitrites cause localised (i.e. pitting) corrosion

[10.15]. Hence, at present, the use of inhibitors is very limited. In fact, it is not explicitly recommended [10.16, 10.17], and sometimes even prohibited [10.18, 10.19]. In any case, it seems that the use of inhibitors must be considered in conjunction with, and not in lieu of, a dense concrete cover of adequate thickness.

In view of the preceding discussion, it may be noted that the nitrites may inhibit corrosion when the presence of chlorides is involved. Some later experimental data indicate that sodium nitrite ($NaNO_2$) may also reduce, or even completely prevent, corrosion due to carbonation. On the other hand, the same nitrite was not effective in preventing corrosion when carbonation took place in chlorides containing concrete [10.20]. The available data, in this respect, are, however, insufficient to warrant practical conclusions.

10.6. OXYGEN PENETRATION

It was shown earlier (eqn (10.3)) that the reactions at the cathode involve the consumption of oxygen. Hence, the corrosion process depends on the presence of oxygen, and the rate of corrosion, on the rate of the oxygen supply at the cathode. The oxygen originates in the air surrounding the concrete, and the amount available at the rebars level depends on the rate of oxygen diffusion through the concrete cover. Hence, similar to the rates of CO_2 and chloride diffusion, the rate of oxygen diffusion depends on porosity of the concrete cover, and decreases, accordingly, with the decrease in the W/C ratio and the efficiency of curing. This effect of the W/C ratio is demonstrated in Fig. 10.17(A), and the expected effect of the cover thickness in Fig. 10.17(B). It is

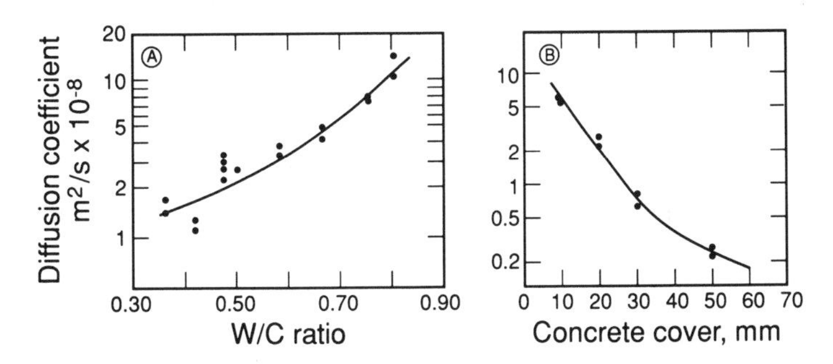

Fig. 10.17. Effect of (A) W/C ratio and (B) specimens thickness on the diffusion coefficient of oxygen at 20°C and 65% RH. Specimens approximately 1 year old. (Adapted from Ref. 10.21.)

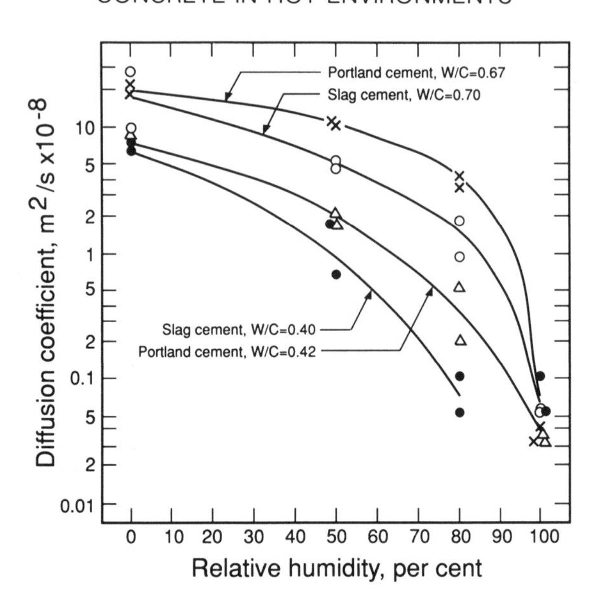

Fig. 10.18. Effect of relative humidity and type of cement on the diffusion coefficient of oxygen at 20°C. Concrete specimens 6–12 months old. Slag content 65%. (Adapted from Ref. 10.21.)

evident that the rate of oxygen diffusion decreases with a decrease in W/C ratio and increase of the cover thickness.

The diffusion of oxygen is very much dependent on the moisture content of the concrete. For the diffusion to take place, the pore system of the cement paste must be, at least partly, dry because the diffusion of oxygen in water is very slow. This effect of moisture content is presented in Fig. 10.18, and it can be seen that, indeed, the coefficient of diffusion decreases with an increase in moisture content (i.e. relative humidity) and particularly when the relative humidity exceeds, say, 80%. It may be noted, as well, that the diffusion coefficient varies in different cements, and that slag cement is characterised by a lower coefficient than OPC.

10.7. EFFECT OF ENVIRONMENTAL FACTORS ON RATE OF CORROSION

The environmental factors which affect the rate of corrosion process are mainly temperature and relative humidity, and these were discussed previously with respect to corrosion-related processes such as carbonation and chloride penetra-

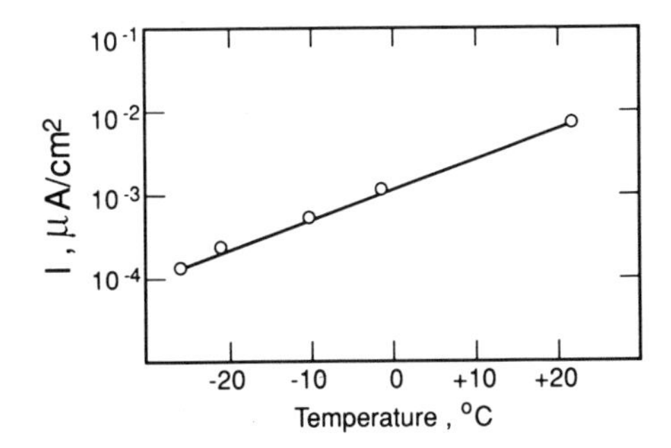

Fig. 10.19. Effect of temperature on corrosion rate at 100% RH (W/C = 0·9, carbonated concrete). (Adapted from Ref. 10.22.)

tion. Generally, it may be expected that environmental factors would affect corrosion in a similar way and this is generally the case when the effect of temperature is considered, i.e. the rate of corrosion increases with the rise in the latter. This effect of temperature is indicated in Fig. 10.19 in which the rate of corrosion is measured by the intensity of the corrosion current.

The effect of relative humidity on the rate of corrosion is presented in Fig. 10.20, in which, again, the corrosion rate is measured by the intensity of the corrosion current. It is apparent that hardly any corrosion takes place when the relative humidity is lower than, say, 85%. In other words, for a significant amount of corrosion to take place the relative humidity must exceed 85%.

Other data, presented in Fig. 10.21, suggest the very same conclusion, namely, that the rate of corrosion becomes significant only when the relative humidity reaches 85%. Moreover, it is also clearly evident from Fig. 20.21 that the effect of temperature on the rate of corrosion is negligible at the lower range of relative humidity. It becomes significant, however, at 85% RH and, indeed, very significant at 95% RH. The preceding conclusions are of practical importance, implying that the risk of corrosion in a hot, dry environment is rather limited and, in this respect, it has been suggested that no corrosion is to be expected when the relative humidity remains below 70% (see Table 10.3). On the other hand, intensive corrosion is to be expected in a hot, wet environment. This may be also the case in marine environment of arid zones

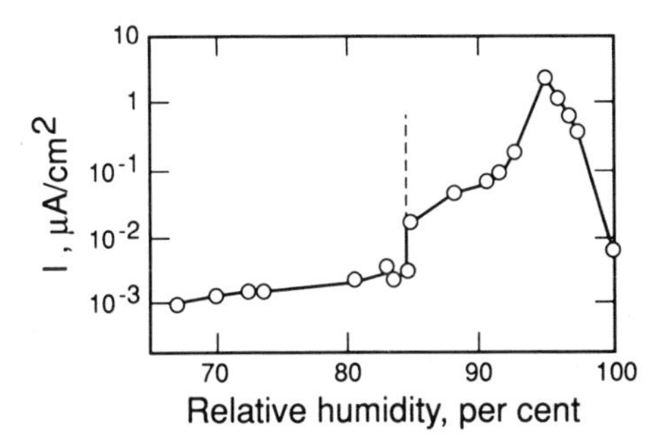

Fig. 10.20. Effect of relative humidity on corrosion rate (W/C = 0·9, carbonated concrete). (Adapted from Ref. 10.22.)

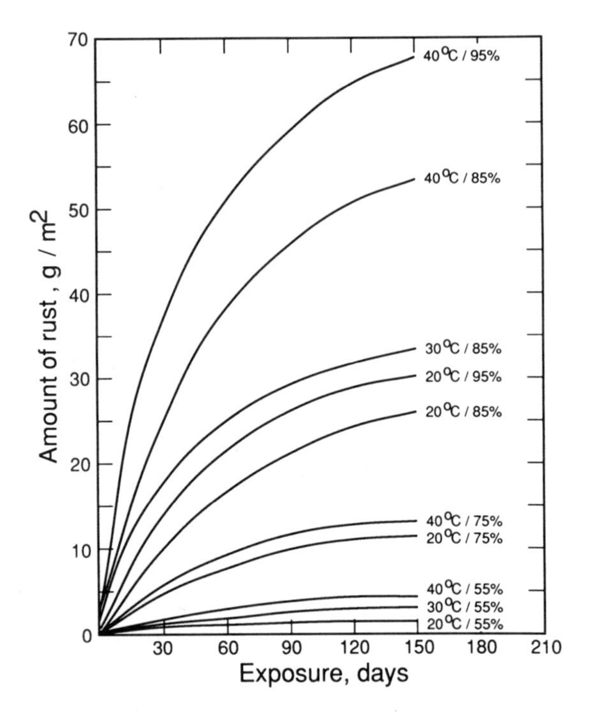

Fig. 10.21. Effect of temperature and relative humidity on rate of corrosion. (Adapted from Ref. 10.23.)

because, in such an environment, the moisture content in the air may be high enough to induce corrosion which is further aggravated by the presence of chlorides.

10.8. EFFECT OF CEMENT TYPE ON RATE OF CORROSION

It was shown earlier (Figs 10.5, 10.6, and 10.9) that the rate of carbonation in concretes made of slag and fly-ash cements is greater than in concretes made of Portland cement. Hence, when no chloride-induced corrosion is to be considered, the latter cement is to be preferred. Otherwise, blended cements are preferable due to their fine pore structure (Chapter 3, Figs 3.3 and 3.15), which slows down considerably the rate of chloride penetration (see section 10.5.1.2). In this respect, slag cement is to be preferred to fly-ash cement because its pore structure is not sensitive to elevated temperatures (Chapter 9,

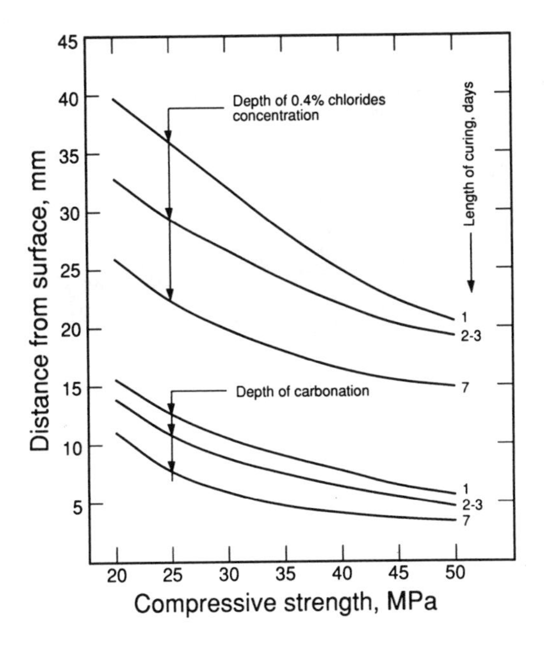

Fig. 10.22. Effect of curing time and concrete strength on depth of carbonation and chloride penetration (0·4% content) at the age of 1 year. (Adapted from Ref. 10.2.)

Fig. 9.3) and, in fact, it even becomes finer with the rise in temperature (Chapter 9, Fig. 9.4).

The preceding conclusion that blended cements, and particularly slag cements, are preferable when chloride-induced corrosion is to be considered, may be questioned because the latter cements are associated with a greater rate of carbonation. That is, it may be asked to what extent, if at all, their adverse effect on the rate of carbonation offsets their beneficial effect on the rate of chloride penetration. This, however, is not the case because the rate of chloride penetration is greater than the rate of carbonation. This is indicated by the data of Fig. 10.22 which compare, at the age of 1 year, the depth of carbonation to the depth at which the chloride content reached the critical concentration of 0·4%, in concretes of the same strength. It follows that the rate of chloride penetration, rather than the rate of carbonation, determines the length of the initiation period and, accordingly, slowing down the rate of chloride penetration must be considered more important than slowing down the rate of carbonation. Hence, it may be concluded that when chloride-induced corrosion is to be considered, the use of blended cements is preferable. Indeed, this latter conclusion is evident from the data presented in Fig. 10.23 in which,

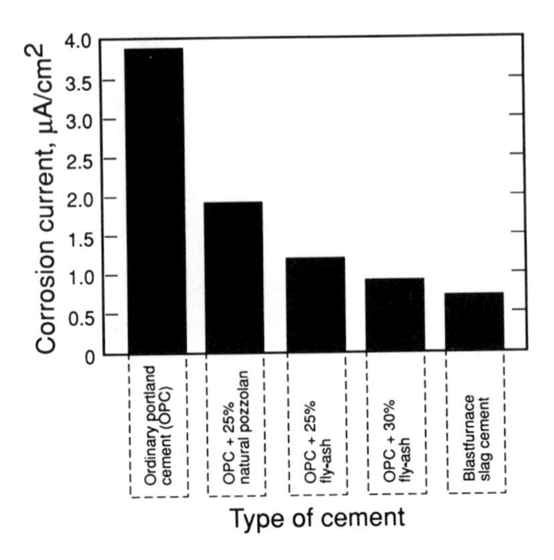

Fig. 10.23. Effect of type of cement on rate of corrosion (concrete specimens, W/C = 0·45, total cement content 375 kg/m^3, immersion in 5% sodium chloride solution). (Adapted from Ref. 10.24.)

again, the corrosion rate is presented by the density of the corrosion current. It may be noted that, in this respect, slag cement gave the best results, followed by fly-ash cements and then by pozzolanic cement.

10.9. PRACTICAL CONCLUSIONS AND RECOMMENDATIONS

Concrete, due to its high alkalinity, provides adequate protection to the rebars against corrosion. Hence, preventing the carbonation front from reaching rebars level may suffice to provide the required protection. When chloride-induced corrosion is to be considered, not only carbonation, but also the chloride ions must be prevented from reaching the rebars level. Furthermore, if the protection provided by the concrete is lost, the corrosion process is still conditional on the presence of water and oxygen. Hence, neither water nor oxygen must be allowed to reach the rebars. To this end, providing the rebars with a dense concrete cover of adequate thickness constitutes the most practical means of providing the required protection. Such protection can be provided also by treating the rebars with an impervious chloride-resistant coating, such as zinc or epoxy coatings. Such treatment, however, is costly and far from

Table 10.1. Nominal Thickness of Cover, Maximum W/C Ratio and Minimum Cement Content to Meet Durability Requirement of Concrete in Accordance with BS 8110, Part 1, 1985[a]

Conditions of exposure[b]	*Nominal thickness of cover (mm)*				
Mild	25	20	20[c]	20[c]	20[c]
Moderate	—	35	30	25	20
Severe	—	—	40	30	25
Very severe	—	—	50[d]	40[d]	30
Extreme	—	—	—	60[d]	50
Maximum free W/C ratio	0·65	0·60	0·55	0·50	0·45
Minimum cement content (kg/m^3)	275	300	325	350	400
Lowest grade of concrete	C30	C35	C40	C45	C50

[a] *This table relates to normal-weight aggregate of 20 mm nominal maximum size.*
[b] *For definitions see Table 10.2.*
[c] *Thickness may be reduced to 15 mm provided that the nominal maximum size of aggregate does not exceed 15 mm.*
[d] *Air entrainment should be used in concrete subjected to freezing whilst wet.*

economically feasible under normal conditions. Hence, such means are not considered further in the text.

The need to provide the rebars with a dense concrete cover manifests itself in specifying a maximum W/C ratio, a minimum cement content, and a minimum thickness of the concrete cover. In the relevant Codes of Practice these requirements are usually related to the expected exposure conditions, and

Table 10.2. Classification of Exposure Conditions in Accordance with BS 8110, Part 1, 1985

Environment	*Exposure conditions*
Mild	Concrete surface protected against weather or aggressive conditions.
Moderate	Concrete surfaces sheltered from severe rain or freezing whilst wet. Concrete subject to condensation. Concrete surfaces continuously under water. Concrete in contact with non-aggressive soil.
Severe	Concrete surfaces exposed to severe rain, alternative wetting and drying or occasional freezing or severe condensation.
Very severe	Concrete surfaces exposed to seawater spray, de-icing salts (directly or indirectly), corrosive fumes or severe freezing conditions whilst wet.
Extreme	Concrete surfaces exposed to abrasive action, e.g. seawater carrying solids or flowing water with pH $\leqslant$ 4·5 or machinery or vehicles.

Table 10.3. Classification of Exposure Conditions.[a]

Environment	*Exposure conditions*
Not aggressive	Indoor conditions with relative humidities always below 70% RH (concrete without chlorides). Constantly and fully immersed structures (low oxygen supply).
Moderately aggressive (without chlorides)	Relative humidity constantly over 70% RH. Changing humidity but with infrequent major variations. Occasional condensation.
Aggressive (without chlorides)	Frequent major variations in humidity, frequent condensation or frequent wetting and drying.
Very aggressive	Severe chloride attack (chloride containing splash water). Frequent wetting and drying. Seawater environment.
Extremely aggressive	Very severe chloride attack, e.g. frequent contact with chloride containing splash water, horizontal surfaces subject to application of chloride-based de-icing salts.

[a]In accordance with Ref. 10.25.

1. Drying periods promote carbonation, whereas wet periods promote corrosion if carbonation has reached the reinforcement. Therefore, the corrosion risk increases with increasing time of the dry periods. That means that climates with long dry and short wet periods may require a higher quality of concrete cover than climates with short dry and long wet periods.

2. The risk of chloride-induced corrosion increases considerably after carbonation of concrete, because initially bound chlorides are released after carbonation and thus increase the amount of free 'corrosive' chlorides.

3. As a rule, all processes involved are accelerated with increasing temperature.

4. When choosing concrete composition, future changes of environmental conditions, resulting from, for example, change in use, should be considered.

the more severe the conditions, the more strict the imposed requirements. British practice, for example, is presented in Table 10.1 in accordance with BS 8110, Part 1, 1985. It may be noted that the standard recognises five conditions of exposure which extend from 'mild' to 'extreme', and are defined in Table 10.2. Somewhat different definitions, which are more corrosion-orientated, are presented in Table 10.3.

It should be realised that the preceding classifications of exposure conditions, and the associated recommendations, do not explicitly consider the effect of temperature, whereas the corrosion processes are affected by the latter. Hence, this effect must be allowed for in a hot environment. This is, perhaps, not that important in a hot, dry environment where the effect of temperature on corrosion is of a limited nature (Fig. 10.21). However, in a hot, wet environment the effect of temperature is rather significant and must be allowed for either by upgrading the quality of the concrete or by increasing the thickness of the cover, or by both (see footnote 1 to Table 10.3).

Adequate curing is all-important in producing dense concrete. The required length of curing depends on many factors such as the setting properties of the cement involved and the environmental conditions. Hence, it is difficult to specify the exact duration of the required time of curing. It is suggested that, in a hot, dry environment, curing for at least 7 days must be considered when OPC is used and, say, 10 days when blended cements are involved. In a hot, wet environment shorter curing periods may suffice because the accelerated effect of temperature on the hydration is not associated with drying.

The type of cement most suitable, from the corrosion point of view, was discussed in some detail in section 10.8. It follows that, under conditions where no chlorides are involved, OPC is preferable to blended cements. However, when chloride-induced corrosion is to be considered, blended cements are more suitable and are, therefore, recommended. In this respect, slag cement, containing 65% slag, seems to exhibit a better performance than its fly-ash or pozzolanic counterparts.

REFERENCES

10.1. Wierig, H.J., Longtime studies on the carbonation of concrete under normal outdoor exposure. In *Proc. RILEM Seminar on Durability of Concrete Structures Under Normal Exposure.* Universitat Hannover, Hannover, 1984, pp. 239–53.

10.2. Jaegermann, C. & Carmel, D., Factors affecting the penetration of chlorides and depth of carbonation. Research Report 1984–1987, Building Research Station, Technion—Israel Institute of Technology, Haifa, Israel, Jan. 1988 (in Hebrew with an English summary).

10.3. Mori, T., Shirayama, K. & Yoda, A., The neutralization of concrete, the corrosion of reinforcing steel and the effects of surface finish. In *Rev. 19th General Meeting.* Cement Association of Japan, Tokyo, Japan, 1965, pp. 249–55.

10.4. Freedman, S., Carbonation treatment of concrete masonry units. *Modern Concrete,* **33**(5) (1969), 33–41.

10.5. Kasai, Y., Matsui, J., Fukushima, Y. & Kamohara, H., Air permeability and carbonisation of blended cement mortars. In *Fly-ash, silica fume, slag and other mineral by-products in concrete* (ACI Spec. Publ. SP 79, Vol. I), ed. V.M. Malhotra. ACI, Detroit, MI, USA, 1983, pp. 435–51.

10.6. Al-Amoudi, O.S.B., Rasheeduzzafar & Maslehuddin, M., Carbonation and corrosion of rebars in salt contaminated OPC/PFA concretes. *Cement Concrete Res.*, **21**(1) (1991), 38–50.

10.7. Jaegermann, C. & Carmel, D., Factors influencing chloride ingress and depth of carbonation. Second Interim Report 1988–1989, National Building Research Institute, Technion—Israel Institute of Technology, Haifa, Israel, April 1989 (in Hebrew with an English summary).

10.8. Rasheeduzzafar, Al-Saadoun, S.S., Al-Gathani, A.S. & Dakhil, F.H., Effect of tricalcium aluminate content on corrosion of reinforcing steel in concrete. *Cement Concrete Res.*, **20**(5) (1990), 723–38.

10.9. Pollock, D.J., Concrete durability tests using the Gulf environment. In *Proc. 1st Intern. Conf. on Deterioration and Repair of Reinforced Concrete in the Arabian Gulf*, Bahrain, 1985, The Bahrain Society of Engineers, Bahrain, Vol. I, pp. 427–41.

10.10. Bakker, R.F.M., On the causes of increased resistance of concrete made of blast furnace slag cement to alkali silica reaction and to sulphate corrosion. Doctorate Thesis, T.H. Aachen, Germany, 1980, (in German).

10.11. Roy, D.M., Kumar, A. & Rhodes, J.P., Diffusion of chloride and cesium ions in Portland cement pastes and mortars containing blastfurnace slag and fly ash. In *Fly Ash, Silica Fume, Slag and Natural Pozzolanas in Concrete* (ACI Spec. Publ. SP 91, Vol. II), ed. V.M. Malhotra. ACI, Detroit, MI, USA, 1986, pp. 1423–45.

10.12. Marusin, S.L., Chloride ion penetration in conventional concrete and concrete containing condensed silica fume. In *Fly Ash, Silica Fume, Slag and Natural Pozzolanas in Concrete* (ACI Spec. Publ. SP 91, Vol. II), ed. V.M. Malhotra. ACI, Detroit, MI, USA, 1986, pp. 1119–31.

10.13. Hanson, C.M., Strunge, H., Markussen, J.B. & Frolund, T., The effect of cement type on the diffusion of chlorides. *Nordic Concrete Res.*, **4** (1985), 70–80.

10.14. Page, C.L., Short, N.R. & El Tarras, A., Diffusion of chloride ions in hardened cement pastes. *Cement Concrete Res.*, **11**(3) (1981), 395–406.

10.15. Ramachandran, V.S., *Concrete Admixtures Handbook*. Noye Publications, Park Ridge, NY, USA, 1984, pp. 540–1.

10.16. RILEM Technical Committee 60-CSC, *Corrosion of Steel in Concrete*, ed. P. Schiessl. Chapman and Hall, NY, USA, 1988.

10.17. ACI Committee 212, Chemical admixtures for concrete. *ACI Mater. J.*, **86**(3) (1989), 297–327.

10.18. South African National Building Research Institute, Interim recommendations by the national research institute to reduce the corrosion of reinforced steel in concrete. *Trans. S. Afr. Inst. Civil Engng* (Johannesburg), **7**(8) (1975), 248–50.

10.19. *Beton-Kalender*, Ernst & Sons, Berlin, Vol. I, 1990, p. 22 (in German).

10.20. Alonso, C. & Andrade, C., Effect of nitrite as a corrosion inhibitor in contaminated and chloride-free carbonated mortars. *ACI Mater. J.*, **87**(2) (1990), 130–7.

10.21. Tuutti, K., *Corrosion of Steel in Concrete*. Swedish Cement and Concrete Res., Inst., Stockholm, Sweden, 1982.

10.22. Tuutti, K., Service life of structures with regard to corrosion of embedded steel. In *Performance of Concrete in Marine Environment* (ACI Spec. Publ. SP-65). ACI, Detroit, MI, USA, 1980, pp. 223–36.

10.23. Raphael, M. & Shalon, R., A study of the influence of climate on corrosion of reinforcement. In *Proc. RILEM 2nd Int. Symp. on Concrete and Reinforced Concrete in Hot Countries*, Haifa, 1971, Vol. I, Building Research Station, Technion—Israel Institute of Technology, Haifa, pp. 77–96.

10.24. Maslehuddin, M., Al-Mana, A.I., Saricimen, H. & Shamim, M., Corrosion of reinforcing steel in concrete containing slag. *Cement Concrete and Aggregates*, **12**(1) (1990), 24–31.

10.25. Schiessl, P. & Bakker, R., *Measures of Protection*. Cited in Ref. 10.16, pp. 71–2.

List of Relevant Standards

BRITISH STANDARDS

BS 12, 1989: *Portland Cements*
BS 146, Part 2, 1973: *Portland Blastfurnace Cement*
BS 1370, 1979: *Low-Heat Portland Cement*
BS 1881: *Methods of Testing Concrete*
Part 102, 1983: *Method for Determination of Slump*
Part 103, 1983: *Method for Determination of Compacting Factor*
Part 104, 1983: *Method for Determination of Vebe Time*
BS 3892: *Pulverised Fuel-Ash*
Part 1, 1982: *Pulverised Fuel-Ash for Use as a Cementitious Component in Structural Concrete*
Part 2, 1984: *Pulverised Fuel-Ash for Use in Grouts and for Miscellaneous Uses in Concrete*
BS 4027, 1980: *Sulphate-Resisting Portland Cement*
BS 4246, Part 2, 1974: *Low-Heat Portland Blastfurnace Cement*
BS 4550: *Methods of Testing Cement*: Part 3, *Physical Tests*
Section 3.1, 1978: *Introduction*
Section 3.2, 1978: *Density*
Section 3.3, 1978: *Fineness*
Section 3.4, 1978: *Strength Tests*
Section 3.5, 1978: *Determination of Standard Consistence*
Section 3.6, 1978: *Test for Setting Times*
Section 3.7, 1978: *Soundness Test*
Section 3.8, 1978: *Test for Heat of Hydration*
BS 6100: *Building and Civil Engineering Terms*: Section 6.1: *Binders*

BS 6588, 1985: *Portland Pulverised-Fuel Ash Cement*
BS 6610, 1985: *Pozzolanic Cement with Pulverised Fuel Ash as Pozzalana*
BS 6699, 1986: *Ground Granulated Blastfurnace Slag for Use with Portland Cement*

ASTM (AMERICAN) STANDARDS

ASTM C109-90: *Test Method for Compressive Strength of Hydraulic Cement Mortars (Using 2 in or 50 mm Cube Specimens)*
ASTM C114-88: *Methods for Chemical Analysis of Hydraulic Cement*
ASTM C125-88: *Definitions of Terms Relating to Concrete and Concrete Aggregates*
ASTM C143-89a: *Test Methods for Slump of Portland Cement Concrete*
ASTM C150-89: *Portland Cement*
ASTM C151-89: *Test Method for Autoclave Expansion of Portland Cement*
ASTM C186-86: *Test Method for Heat of Hydration of Hydraulic Cement*
ASTM C191-82: *Test Method for Time of Setting of Hydraulic Cement by Vicat Needle*
ASTM C204-89: *Test Method for Fineness of Portland Cement by Air Permeability Apparatus*
ASTM C219-90: *Standard Terminology Relating to Hydraulic Cement*
ASTM C260-86: *Air-Entraining Admixtures for Concrete*
ASTM C403-88: *Test Method for Time of Setting of Concrete Mixtures by Penetration Resistance*
ASTM C452-89: *Test Method for Potential Expansion of Portland Cement Mortars Exposed to Sulfate*
ASTM C494-86: *Chemical Admixtures for Concrete*
ASTM C595-89: *Blended Hydraulic Cements*
ASTM C618-89a: *Fly Ash and Raw or Calcined Natural Pozzolan for Use as a Mineral Admixture in Portland Cement Concrete*
ASTM C989-89: *Ground Iron Blast-Furnace Slag for Use in Concrete and Mortars*
ASTM 1017-85: *Chemical Admixtures for Use in Producing Flowing Concrete*

Selected Bibliography

PROCEEDINGS OF SYMPOSIA

1. *Proc. RILEM Int. Symp. on Concrete and Reinforced Concrete in Hot Countries*, Haifa, 17–19 July, 1960 (Vols 1 and 2). Building Research Station, Technion —Israel Institute of Technology, Haifa.
2. *Proc. RILEM 2nd Int. Symp. on Concrete and Reinforced Concrete in Hot Countries*, Haifa, Aug. 2–5, 1971 (Vols I and II). Building Research Station, Technion—Israel Institute of Technology, Haifa.
3. *Proc. 3rd RILEM Conf. on Concrete in Hot Climates*, Torquay, U.K., 21–25 Sept., ed. M. Walker, F.N. Spon Ltd, London, 1992.
4. *Proc. Int. Seminar on Concrete in Hot Countries.* Helsingor, Sweden, 1981, Skanska, Malmo, Sweden.
5. *Proc. 1st Int. Conf. on Deterioration and Repair of Reinforced Concrete in the Arabian Gulf*, Bahrain, 26–29 Oct. 1985 (Vols I and II). The Bahrain Society of Engineers, Manama, Bahrain.
6. *Proc. 2nd Int. Conf. on Deterioration and Repair of Reinforced Concrete in the Arabian Gulf*, Bahrain, 11–13 Oct. 1987 (Vols I and II). The Bahrain Society of Engineers, Manama, Bahrain.
7. *Proc. 3rd Int. Conf. on Deterioration and Repair of Reinforced Concrete in the Arabian Gulf*, Bahrain, 21–24 Oct. 1989 (Vols I and II). The Bahrain Society of Engineers, Manama, Bahrain.

GUIDES

1. STUVO (the Dutch member group of FIP), *Concrete in Hot Countries.* S-Hertogenbosch, The Netherlands.
2. *The CIRIA Guide to Concrete Construction in the Gulf Region.* CIRIA Spec. Publ. 31, London, U.K., 1984.
3. *Concrete Construction in Hot Weather.* FIP Guide to good practice, Thomas Telford, London, U.K., 1986 (reprinted 1989).
4. *Concrete in Warm Climate—Current Knowledge and Recommendations.* Annales de l'Institute Technique du Batiment et des Travaux Publics, No. 474 (Beton 265), May 1989 (in French), pp. 77–119.
5. ACI Committee 305, Hot weather concreting. *ACI Mater. J.*, **88**(4) (1991), 417–36.

REVIEWS AND BIBLIOGRAPHIES

1. Shalon, R., Report on behaviour of concrete in hot climate, Part I. *Mater. Struct.*, **11**(62) (1978), 127–31.
2. Shalon, R., Report on behaviour of concrete in hot climate, Part II: Hardened concrete. *Mater. Struct.*, **13**(76) (1980), 255–64.
3. Samarai, M., Popovics, S. & Malhotra, V.M., Effects of high temperatures on the properties of fresh concrete. *Transp. Res. Rec.*, **924** (1983), 42–50.
4. Samarai, M., Popovics, S. & Malhotra, V.M., Effects of high temperatures on the properties of hardened concrete. *Transp. Res. Rec.*, **924** (1983), 50–6.
5. Samarai, M., Popovics, S. & Malhotra, V.M., Effects of high temperatures on the properties of fresh and hardened concrete: A bibliography (1915–1983). *Transp. Res. Rec.*, **924** (1983), 56–63.
6. Ali, M.A., Concrete in hot climates—A literature review of temperature effects on the properties and performance of concrete. British Research Establishment Note, Sept. 1986.
7. Bibliography in *Hot Weather Concreting.* British Cement Association, Concrete Information Service, May 1990.

Author Index

Reference numbers refer to the publication of the author in question and the numbers in parentheses to the pages on which the references appear in the text. Figure numbers refer to the figures reproduced from the author's publication or based on his data. The numbers in parentheses are the pages on which the figures appear in the text.

Subject Index